Technician Safety and Laboratory Practice

This ... at
level ... e.
This ... d
altho ... ll
level ... ly
follo ... o-
grap ... nt
tests

Dr. ... rd
Coll ... y,
Phys ... nd
Bioc ... of
the ... d
man ... e-
rienc ... ic
jour

Technician Safety and Laboratory Practice

M.D. Hawkins
M.A., M.Sc., Ph.D., C.Chem., F.R.I.C.
Head of Physical Sciences, Bedford College of Higher Education

Cassell • London

CASSELL LTD.
35 Red Lion Square, London WC1R 4SG
and at Sydney, Auckland, Toronto, Johannesburg,

an affiliate of Macmillan Publishing Co. Inc.,
New York

First published 1980

I.S.B.N. 0 304 30550 2

Line drawings by Oxford Illustrators Ltd.

Colset Pte. Ltd.
Singapore

Printed and bound in Great Britain at
The Camelot Press Ltd, Southampton

Preface

This book was written to cover the content of the Level 1 Safety and Laboratory Practice unit (U76/001) for Technician Certificate courses of the Technician Education Council (TEC). The text follows the order and numbering of the general and specific objectives of the standard unit, but the content has been extended slightly to include a brief account of the hazards of biological and medical laboratories, the dangers of compressed gases and of evacuated glassware and a number of other important topics. The specific objectives are listed at the beginning of each section so the student knows precisely what he is expected to achieve.

I am aware of the difficulties of both the teacher and the day-release student in covering five (and possibly six) TEC units in a year on the student's single day each week at college. The book is comprehensive and self contained. It aims to give a sufficiently detailed treatment so that the time devoted to note taking will be minimised and an absentee can easily catch up the work he has missed. A greater proportion of the time available can then be devoted to practical work, demonstration and discussion. Allowance has been made for the fact that the unit may be taught separately or that much of its content may be integrated into other units of the TEC programme. About sixty assignments are included which the student can carry out in the laboratory or at work. A number of objectives, for example those included in the sections on measuring instruments, are best achieved by practical work and the relevant sections and assignments have been planned with this in mind. Other assignments were designed with the intention of making the student aware of the hazards of his working environment and of instilling a greater sense of 'safety consciousness'. Over two hundred short answer questions and a seventy-five item objective test provide the means of assessing whether the course objectives have been achieved. The short answer questions can be used for homework and to provide the basis for class discussion or as part of the in-course assessment programme.

I should like to thank those whose encouragement, help and example have made the book possible. Particular thanks are due to Dr Lester Crook and Miss Fiona Foley of Cassell's; Messrs B.H. Jones, F.A. Milemore, K. Squire, F. Stafford, C.L. Wale, A.G. Watt, Mrs E. Kent and other colleagues; and to the following firms for providing photographs of their products and permission to reproduce copyright material: Association of Science Education, Avo Ltd., B.D.H. Chemicals, Chubb Ltd., Griffin and George, Philip Harris, Hopkin and Williams, L. and G. Fire Appliance Co. Ltd., Moore and Wright Ltd., Nikon Microscopes, Oertling, Olympus Microscopes, Starrett and Co. Ltd.,

Preface

St. John's Ambulance and the British Red Cross Society. I should also like to thank Mr R.F. Harburn for taking a number of excellent photographs at very short notice.

Hazard warning symbols appearing on the cover, and within the text are Crown copyright, reproduced with the permission of the Controller of Her Majesty's Stationery Office.

December 1979 M.D.H.

Introduction

Any laboratory can appear to be a dangerous and bewildering place to a school-leaver beginning his (or her) career as a laboratory technician. The wide use of instrumental methods has changed, but not lessened, the skills required; nor has it reduced the great importance of laboratory safety. In fact, the common application of highly toxic, inflammable or corrosive substances and the widespread use of electrical equipment, micro-organisms, radioactive materials, lasers and other potentially dangerous techniques in the laboratories of schools, colleges, universities, hospitals, industry and research establishments has enormously increased the range of possible hazards. In addition to learning how to use specialised equipment and instruments, a technician needs to know how to carry out such basic operations as weighing and the measurement of volumes, length, current and voltage. These and other important information and skills are described in the following chapters on laboratory safety, electrical hazards, the storage and use of dangerous materials, fire prevention, the law and technicians, scientific reporting and first aid.

Proficiency in practical skills comes only with training, experience and knowledge. Although the importance of these factors should not be underestimated, it must be stressed that safety is not a subject which can be learned or taught. Of course, it is essential to be fully aware of the hazards involved and to wear the protective clothing provided and observe the safety procedures recommended at all times, but it is also important to develop an awareness of what constitutes safe and unsafe practice. This awareness can be developed in even routine laboratory operations, such as transferring inflammable liquids, pouring concentrated acids and evacuating glassware, by continually asking oneself the question 'What would happen if ...?'. For example, 'What would happen if someone struck a match to light a Bunsen burner or a cigarette while I'm doing this?'; 'What would happen if I dropped the bottle or if the flask breaks?'; 'What would happen if a fire started in this storeroom?'; 'What action would I take and how would I get out?'. The answers to these questions and the steps taken to minimise the hazards involved provide a valuable lesson in acquiring the correct attitude to laboratory safety that is often referred to as 'safety consciousness'.

Impetus to achieve better standards of safety was provided by the Health and Safety at Work Act, 1974 and by a general insistence on improved working conditions. It is often said that 'accidents do not happen, they are caused': caused that is by such factors as carelessness, tiredness, ignorance, practical jokes, defective equipment, taking risks and short cuts, and by the unwillingness of some employers to provide really effective safety equipment if this costs a little more. But whatever the cause, accidents can be avoided. I hope that the readers of this book will use the knowledge they gain to reduce the enormous number of accidents and injuries which occur every year.

Contents

Preface
Introduction

A Laboratory safety 1

1 *General learning objective*: To know the potential hazards associated with electrically powered equipment and its connection to a supply 1

Specific objectives 1

Introduction 1

1.1 The implications of colour blindness 3

1.2 The BS colour coding for cable and flex 4

1.3 Correct connection of a 13 A plug top to a flex end 5

1.4 The use of the relationship $I = W/V$ to determine the current drawn by an appliance 7

1.5 Selection of the appropriate size of fuse for a stated application 8

1.6 Selection of appropriately rated flex for a specified piece of equipment 12

1.7 Conduction of electricity by the human body 13

1.8 The physiological consequences of the passage of an electric current through the body 14

1.9 Earthing electrical equipment 14

1.10 Reason for earthing a metal clad electrical appliance 15

1.11 Potential accidents in the laboratory resulting from the careless routing of equipment flex 16

1.12 Damaged or worn sockets, plug tops and leads 16

1.13 Danger of trying to draw a current through a cable or flex in excess of its rated value 16

1.14 Use of multiple adaptors and distribution boards 17

Questions: **1** Hazards of electrical equipment 18

2. *General learning objective*: To state the elementary safety precautions which should be borne in mind when using power tools 20

Specific objectives 20

Introduction 20

2.1 Appropriate protective clothing 21

2.2 Hazards of loose clothing 21

2.3 The hazards associated with long hair and any power tool 21

Questions: **2** Using power tools 22

3 *General learning objectives*: To recognise the elementary hazards which may be encountered in a chemical laboratory 23

Specific objectives 23
Introduction 23
3.1 Fume cupboards 27
3.2 Storage of poisons 29
3.3 Use of reference material to determine chemical hazards 31
3.4 Storage of inflammable liquids 35
3.5 Transport of bulk chemicals from store 36
3.6 The need for always labelling chemical containers 38
3.7 Hygiene 41
3.8 The hazards associated with damaged glassware, cutting tools etc. 41
3.9 Common carcinogens 45
3.10 Precautions if carcinogens do have to be employed 46
Questions: **3** Hazards of chemical laboratories 47

4 *General learning objectives*: To recognise elementary hazards which may be encountered in (a) a physical laboratory and (b) a biological or medical laboratory 48
Specific objectives 48
4.1 Radioactive sources 48
4.2 Ultra-violet sources 53
4.3 Protection against ultra-violet radiation 53
4.4 The potential hazard of laser light 54
4.5 Safety in biological laboratories 55
4.6 Sterilisation of apparatus for microbiological experiments 56
Questions: **4** Hazards of physical and biological or medical laboratories 57
5 *General learning objectives*: To know the basic precautions to prevent fire and the action to be taken in the event of fire 58
Specific objectives 58
5.1 Common fire hazards 58
5.2 Fire doors 63
5.3 The main types of fire extinguisher 64
5.4 Range of application of the main types of fire extinguisher 69
5.5 Use of fire extinguishers 71
5.6 Position of fire alarm buttons 75
5.7 Escape routes and assembly points 75
5.8 Localisation of fire and smoke 75
Questions: **5** Fire 76

6 *General learning objective*: To recognise approved international warning signs 78
Specific objectives 78
Introduction 78
6.1 International sign for radiation hazards 79
6.2 International sign for high voltage 80

6.3 Chemical and biological warning signs 81
Questions: **6** International warning signs 81

7 *General learning objectives*: To know and understand the necessity of maintaining a personal code of safety in the laboratory 84
Specific objectives 84
Introduction 84
7.1 Protective clothing 85
7.2 The necessity for a mature code of behaviour in a laboratory 86
7.3 The principle behind the establishment of such a code 87
7.4 The hazards from smoking, eating and drinking in a laboratory 88
7.5 The hazards of storing food or drink in a laboratory 88
7.6 Sequence of actions when going off duty or after handling dangerous substances 89
7.7 The correct method of lifting heavy objects 90
Questions: **7** Personal code of safety 94

B The law and technicians 95
8 *General learning objectives*: To show an awareness of the major provisions of national law and local regulations as they apply to the technician 95
Specific objectives 95
Introduction 95
8.1 Disposal of dangerous substances 96
8.2 Regulations for handling dangerous substances 98
8.3 The Health and Safety at Work Act 1974 99
Questions: **8** Law 101

C Laboratory practice and equipment handling 102
9 *General learning objectives*: To know the simple techniques of reading instruments and to state their limitations and accuracy 102
Specific objectives 102
Introduction 102
Accuracy and precision 103
9.1 The vernier 103
9.2 The micrometer 107
9.3 Accuracy of measurements made with a vernier and micrometer 109
9.4 Balances and their use 112
9.5 Accuracy of balances 115
9.6 Checking of balances 116
9.7 Use of a burette to deliver a stated volume of liquid 118
9.8 Use of a pipette to deliver a stated volume of liquid 121

	9.9	Accuracy of volumetric glassware	123
	9.10	The multimeter	124
	9.11	The optical microscope	128
	9.12	Correct procedure for cleaning a microscope	133
	9.13	Primary standards for physical, chemical and biological quantities	134
	9.14	Secondary standards	134
	9.15	Periodic testing of common laboratory equipment against standards	138
	9.16	Purpose of the BSI	140
	9.17	Titles of useful BSI publications	141
Questions:	**9**	Instruments	142
D		Scientific reporting	145
10		*General learning objectives*: To understand the necessity for clear, accurate, honest reporting of experimental results	145
		Specific objectives	145
	10.1	The practical notebook	145
	10.2	Reporting of experimental results	148
	10.3	Presentation of experimental results	149
Questions:	**10**	Scientific reporting	159
E		Laboratory skills	162
11		*General learning objectives*: To be able to care for the items of equipment common to all laboratories	162
		Specific objectives	162
	11.1	Cleaning glassware	162
	11.2	Selection of appropriate cleaning method	163
	11.3	Breakages	164
	11.4	Secondary cells	164
	11.5	Charging common laboratory batteries	170
Questions:	**11**	Laboratory skills	172
F		First-aid	173
12		*General learning objectives*: To be able to deal, in the first instance, with a variety of laboratory accidents	173
		Specific objectives	173
		Introduction and definition of first-aid	174
	12.1	Siting of first-aid boxes, eye-wash bottles and first-aid points	176
	12.2	Minimum contents of a first-aid box for laboratory use	177
	12.3	Priorities in administering first-aid	179
	12.4	The purposes of artificial respiration	180
	12.5	Artificial respiration	182
	12.6	First-aid dressings for surface wounds	186
	12.7	Methods of controlling bleeding from a wound	190
	12.8	Location of the main pressure points	192

12.9	Use of a pressure point	192
12.10	Procedures for general treatment of poisons	194
12.11	Procedures for first-aid treatment of thermal burns	196
12.12	Procedures for first-aid treatment of chemical burns	198
12.13	The common antidotes for chemical burns by acids, alkalis, phenol, bromine and phosphorus	198
12.14	The use of eye-wash bottles	199
12.15	First-aid treatment for electric shock	201
12.16	First-aid treatment of wound shock	202
12.17	Unconsciousness	203
12.18	First-aid treatment for factures	204
12.19	The need for reporting laboratory accidents	206
12.20	Reporting procedure for accidents	206
Questions:	**12** First-aid	207

Further reading		209
Objective test		211
Appendixes		220
Appendix I	SI units	220
	Prefixes for multiples and sub-multiples	221
Appendix II	Important physical contents	222
Appendix III	Conversion factors	222
Appendix IV	Relative atomic masses	223
Appendix V	IUPAC nomenclature	224
Appendix VI	Hazchem code	225
Appendix VII	Conductivity of aqueous potassium chloride solutions	226
Appendix VIII	Primary standards for volumetric analysis	227
Appendix IX	Optical density of aqueous potassium chromate (VI) solution	227
Appendix X	Density of pure water	228
Appendix XI	Laboratory emergency information sheet	229
Logarithms		230
Answers to questions		232
Answers to objective test		233
Index		235

A Laboratory safety

Section 1: *The expected learning outcome of this section is that the student should know the potential hazards associated with electrically powered equipment and its connection to a supply*

Specific objectives: *The expected learning outcome is that the student:*

1.1 *Recognises the implications of colour blindness.*
1.2 *States BS colour coding for both cable and flex.*
1.3 *Correctly connects a 13 A plug top to a flex end.*
1.4 *Uses I = W/V to determine the current drawn by an appliance.*
1.5 *Selects the appropriate size of fuse for a stated application.*
1.6 *Uses suppliers' catalogues (or other sources) to select an appropriately rated flex for a specific piece of equipment.*
1.7 *Recognises that the human body can conduct electricity.*
1.8 *Recognises the simple physiological consequences of the passage of an electric current through the body.*
1.9 *Demonstrates how a piece of equipment could be earthed.*
1.10 *States the reason for earthing a metal clad appliance.*
1.11 *Recognises in a laboratory where the careless routing of equipment flex could lead to an accident.*
1.12 *Recognises when plug tops, sockets and leads are damaged or worn and takes the appropriate action.*
1.13 *States the danger of trying to draw a current through a cable or flex in excess of its rated value.*
1.14 *Recognises that care should be exercised in the use of multiple adaptors and distribution boards.*

Introduction

(a) *Units*

The four basic quantities in electricity are current, potential, resistance and power. The corresponding SI units and their symbols are listed in Table 1.1.

1. *Current* An electric current is the name given to the movement of electrons through a conductor when a voltage (or potential difference) is applied between its ends. This current may be used to heat a wire or—as in the case of the white hot filament of an electric light bulb—to produce light. An electric current may also be used to power an electromagnet, to drive a motor or to produce chemical changes, such as those which occur when a battery is recharged.

Table 1.1 *Electrical units*

Quantity	SI unit	Symbol	
Electric current	ampere (or amp)	A	
Potential	volt	V	
Resistance	ohm	Ω	(Greek capital omega)
Power	watt	W	

2. *Potential* Potential or voltage is a measure of the driving force of an electric current. A current flows from a point of high potential to one of lower potential and may be compared with the flow of water downhill. A difference in potential is essential if a current is to flow. The potential of the neutral terminal in a mains socket is zero and is connected to earth at the power station, while that of the live terminal differs from country to country and is usually 240 V in the United Kingdom. The potential difference between the terminals is thus 240 V. The relative positions of the terminals in the three-pin electrical socket outlets commonly used in Britain are shown in fig. 1.1.

3. *Resistance* The resistance of a material is a measure of the difficulty of passing an electric current through it. The electrons in a metal are free to move and a current will flow through a wire or rod of the material as soon as a potential difference is applied between its ends. The metal thus has a low resistance or a high conductivity, as

$$\text{conductivity } (\Omega^{-1} \text{ or siemens, S}) = \frac{1}{\text{resistance } (\Omega)}$$

In most non-metals (except graphite, which is a good conductor) the electrons are not able to move about freely and these materials have an extremely high electrical resistance and are known as non-conductors or insulators. Examples of such materials include wood, rubber, plastic, porcelain and paper.

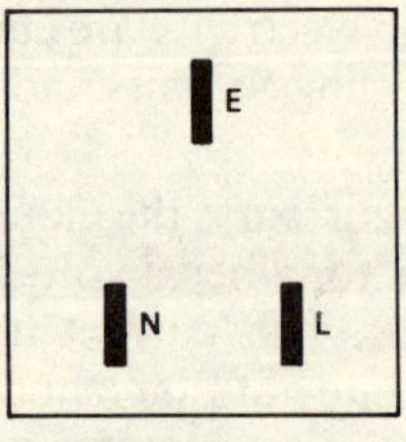

Fig.1.1 Relative positions of terminals in a three-pin socket outlet (front view)

The relationship between the three quantities is described by Ohm's law, which may be expressed as:

$$\frac{\text{potential difference } V}{\text{current } I} = \text{resistance } R$$

or

$$V = I \times R$$

4. *Power* The power rating of a piece of electrical equipment is a measure of the rate of energy transfer, i.e. it is the total amount of energy passing through the equipment in each second (see 1.4).

(b) *Electrical safety*

Many laboratory and workshop operations would be tedious or difficult without electrical power, while others—including the use of many common laboratory instruments and virtually all specialised equipment—would be impossible. However, the ready availability of such a powerful power source presents a number of potential hazards and its misuse can cause serious injury and death (see 1.8). Short circuits and other electrical faults are also a common cause of fires. These dangers must never be underestimated. Most electrical accidents are caused by worn-out equipment or faulty workmanship—both of which are avoidable. All electrical equipment operating at mains voltage can be lethal and thus a potential killer is present in every home or laboratory. The need for continual care is emphasised by the fact that of the six different ways of connecting the three wires of a mains lead to the three terminals of a plug top, only one is correct, while two of the incorrect methods can be lethal even before the equipment is switched on. The methods of avoiding such accidents and of minimising the hazards associated with electrically powered equipment are described in this section.

1.1
The implications of colour blindness

Colour codings are now widely employed for many of the components used in electrical equipment and appliances. The resistance and tolerance values of fixed resistors in electronics, for example, are frequently indicated by a series of coloured bands or rings round the component (see Table 1.2), while a different coloured insulating material is employed to label the live, neutral and earth leads in the flex or cable which is used to connect the equipment to the electrical supply. It is important to observe these colour codes as any mistakes can be expensive in terms of damage to the equipment or dangerous (or even fatal) to the user from electrocution or fire which may result.

A small proportion of the population is unable to distinguish certain colours, particularly red and green. A simple colour test at the beginning of a technician's training and career is therefore useful to identify a deficiency of which the person concerned may be unaware.

Table 1.2 *Colour coding of fixed resistors*

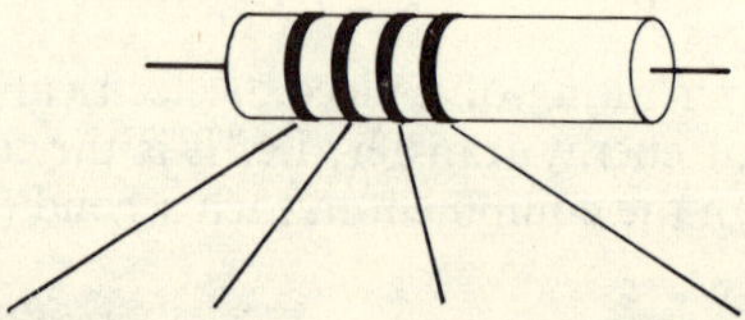

Band colour	First band	Second band	Third band	Fourth band	Tolerance (/%)
Black	0	0	$\times 10^0 = 1$	None	± 20
Brown	1	1	$\times 10^1 = 10$	Silver	± 10
Red	2	2	$\times 10^2 = 100$	Gold	± 5
Orange	3	3	$\times 10^3$	Red	± 2
Yellow	4	4	$\times 10^4$		
Green	5	5	$\times 10^5$		
Blue	6	6	$\times 10^6$		
Violet	7	7			
Grey	8	8			
White	9	9			

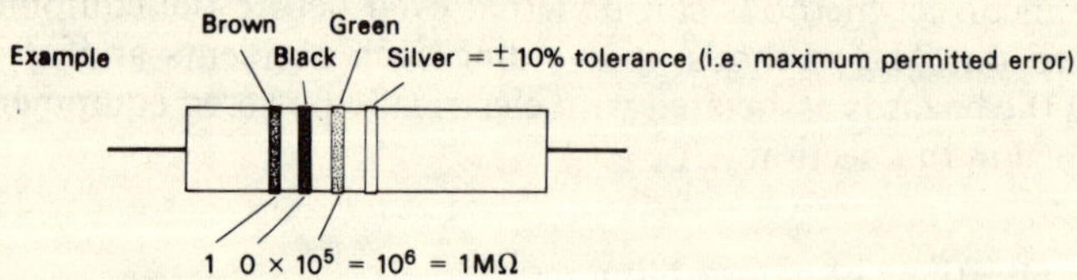

The implications of colour blindness are obvious. Anyone unable to distinguish the different colours employed (or is ignorant of the colour code which applies) should leave the connection of such equipment to a competent electrician.

1.2
The BS colour coding for cable and flex

From 1 July 1970 Great Britain and the majority of European countries adopted the following colour code for cable and flex:

earth	green and yellow stripes	(stripes)
live	brown	(dark)
neutral	blue	(light)

One advantage of this new British Standards Institution code is that the colours are distinguishable by people who are red – green colour blind. The appearance of the different coloured leads to a colour blind person is indicated in parentheses.

Prior to July 1970 British, American and Continental colour codes were all different. Care should therefore be taken in reconnecting equipment bought before that date. The earlier colour code for cable and flex in the United Kingdom was red (live), black (neutral) and green (earth).

1.3
Correct connection of a 13 A plug top to a flex end

Equipment required

1 small screwdriver;
1 pair of wire strippers (see fig. 1.2); 1 13 A plug;
3 A and 13 A cartridge fuses;
1 length of three-core flex.

Method

1. Unscrew the plug top and remove the cartridge fuse. Remove one of the flex clamp retaining screws and loosen the other (see fig. 1.3(a)).
2. Cut away about 50 mm of the *outer* sheath of the flex, taking care not to damage the insulation on the three separate wires and fasten the

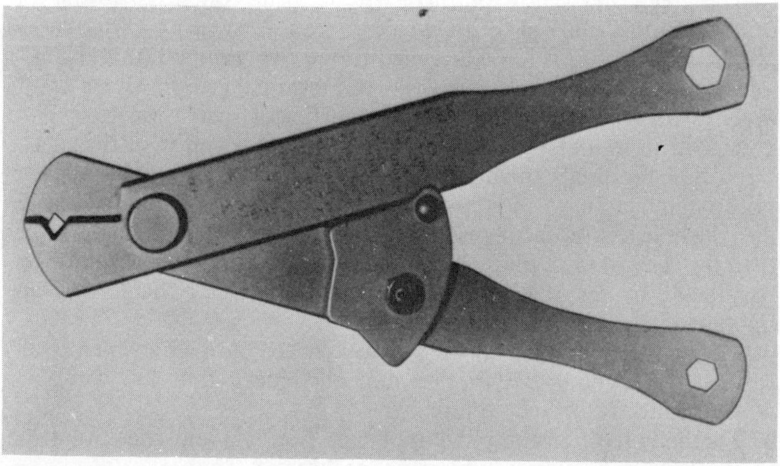

Fig.1.2 Wire strippers

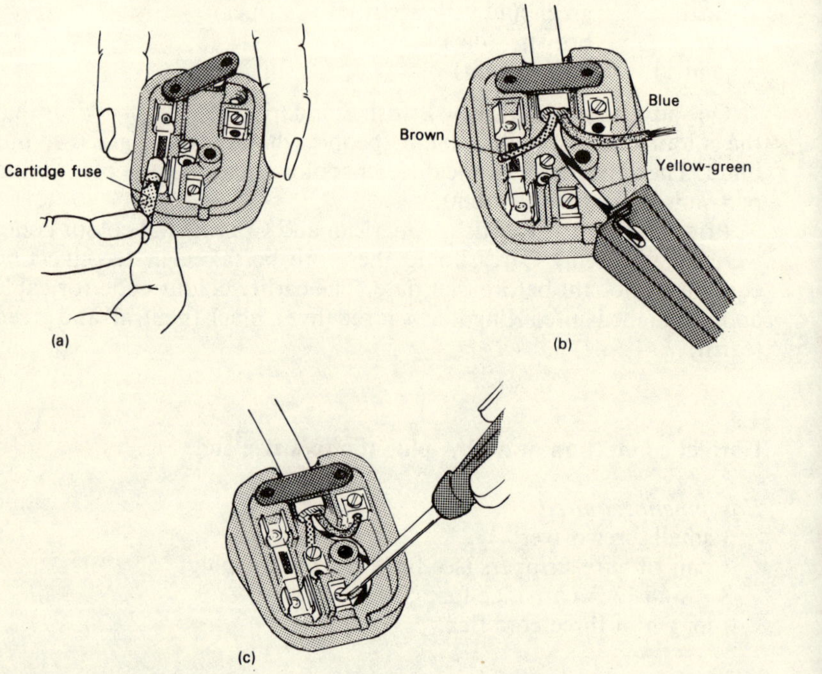

Fig.1.3 Connection of a 13 A plug top to a flex end

sheath firmly under the clamp so that no strain can be placed on the terminals in the plug. Cut the wires to reach approx. 12 mm beyond the appropriate terminal:

The green and yellow or *earth* wire is to be connected to the top centre terminal marked 'E' or ⊥.

The brown wire is to be attached to the *live* terminal (marked 'L') and the blue wire to the *neutral* (or 'N') terminal (see fig. 1.3(b)).

3. Use the wire stripper to remove sufficient insulation to expose enough wire to make the connections (approx. 6 mm is sufficient for screw hole terminals and about 13 mm for the clamp type) taking care not to damage the wires. Twist the strands of each wire in turn and fit into the hole *or* loop clockwise round the terminal and tighten the screw (see fig. 1.3(c)). Looping the strands of wire clockwise around the terminals ensures that they will not loosen when the screws are tightened.
4. Check that there are no stray 'whiskers' of bare wire which could cause short circuits within the plug and check that the wires are connected to the correct terminals.
5. Fit the correct fuse for the appliance (see 1.5) and check that the screws holding the wires in place are secure.
6. Refit the plug top.

Assignment

Connect a flex end to a 13 A plug following the directions given above.

1.4
The use of the relationship I = W/V to determine the current drawn by an appliance

Power is the rate of energy transfer of a machine or electrical appliance.
The SI unit of power is the watt (symbol W):

1 watt = 1 joule per second

where watts = volts (V) × current (I)

i.e. $W = V \times I$.

watts = volts × amps

The current drawn by an appliance can be calculated from

I = watts/volts = W/V.

The following multiples and submultiples of power are commonly employed:

1 milliwatt (1 mW)	= 0.001 watt	= 10^{-3} W
1 kilowatt (1 kW)	= 1000 watts	= 10^{3} W
1 megawatt (1 MW)	= 1 000 000 watts	= 10^{6} W

Example 1.1
Calculate the current drawn by (a) a 100 W bulb and (b) a 2 kW heater operating on a mains voltage of 240 V.

Solution

(a) Current drawn by a 100 W bulb

$$I = W/V = 100/240 = 0.4 \text{ A.}$$

(b) Current drawn by a 2 kW heater

$$I = W/V = 2000/240 = 8.3 \text{ A.}$$

Ohm's law may be used to show the relationship between the electrical power developed in a circuit and the circuit's resistance:

$$\text{power} = V \times I \text{ watts.} \qquad (1)$$

Ohm's law states that

$$V = I \times R \qquad (2)$$

or

$$I = V/R.$$

Thus by substituting $I \times R$ for V in equation (1),

$$\text{power} = (I \times R) \times I = I^2R \text{ watts}$$

and, by substituting V/R for I in equation (2),

$$\text{power} = V \times V/R = V^2/R \text{ watts.}$$

Mechanical power is also measured in watts; nevertheless the term 'horsepower' (hp) is still widely employed where,

$$1 \text{ hp} = 745.7 \text{ W}$$

The watt second or joule is the unit of energy. This unit is too small for most practical purposes and a larger unit, known as the *kilowatt hour* (kW h), has been adopted by the UK Board of Trade for calculating energy costs. A kilowatt hour is the energy absorbed or transformed when 1000 W have been supplied to a circuit for one hour. Thus, for the cost of one Board of Trade electrical unit it is possible to operate a 100 W bulb for 10 h or a 2 kW heater for 0.5 h.

$$1 \text{ kW h} = 3.6 \text{ MJ.}$$

The kilowatt hour is not an SI unit.

Example 1.2
Calculate the weekly energy cost of operating two electric motors for twelve hours each day at 3.2p per unit if the potential difference at the terminals is 440 V and each motor takes 20 A.

Solution
Power of each motor $= V \times I = 440 \times 20 = 8.8$ kW.
Energy consumption of two 8.8 kW motors in $7 \times 12 = 84$ h
$= 2 \times 8.8 \times 84 = 1478.4$ kW h.
Weekly energy cost $= 1478.4 \times 3.2/100 =$ £47.31.

1.5
Selection of the appropriate size of fuse for a stated application

It is important that the plug used to connect electrical equipment to the power supply is fitted with the correct fuse. Complex equipment often contains a number of fuses in different parts of the circuit to protect the various components from damage by ensuring that a current exceeding a certain value cannot pass through them. An excessive current passing through a circuit lead generates large amounts of heat and can therefore become a fire hazard. A nail or length of thick wire must never be used to replace a blown fuse, even as a temporary measure.

The use of 13 A socket outlets fitted with fused plugs has the advantage that a piece of equipment can be plugged into any socket in a mains circuit and still retain its own correct protecting fuse. This fuse is rated to suit the appliance connected to the plug.

The correct fuse for a particular application is one which will allow the safe passage of the current drawn by the appliance when it is functioning correctly as calculated in sub-section 1.4, but which will

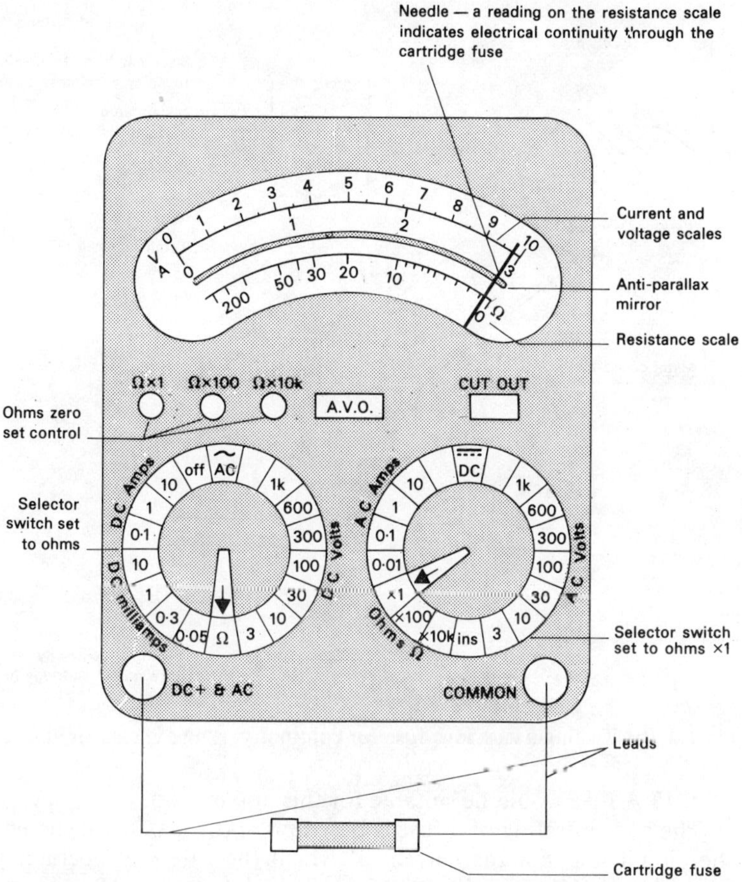

Fig.1.4 (a) Testing a cartridge fuse for continuity using a multimeter

'blow' and thus prevent the current passing if—as a result of a short circuit or other fault developing in the wiring—it should greatly exceed this value. The fitting of an automatic excess current circuitbreaker is a valuable safety precaution.

Example 1.3
Select an appropriate fuse for a 2.5 kW immersion heater for a thermostat bath operating from a 240 V supply.

Solution
Current drawn by a 2.5 kW appliance

$$= W/V = 2500/240 = 10.4 \text{ A}.$$

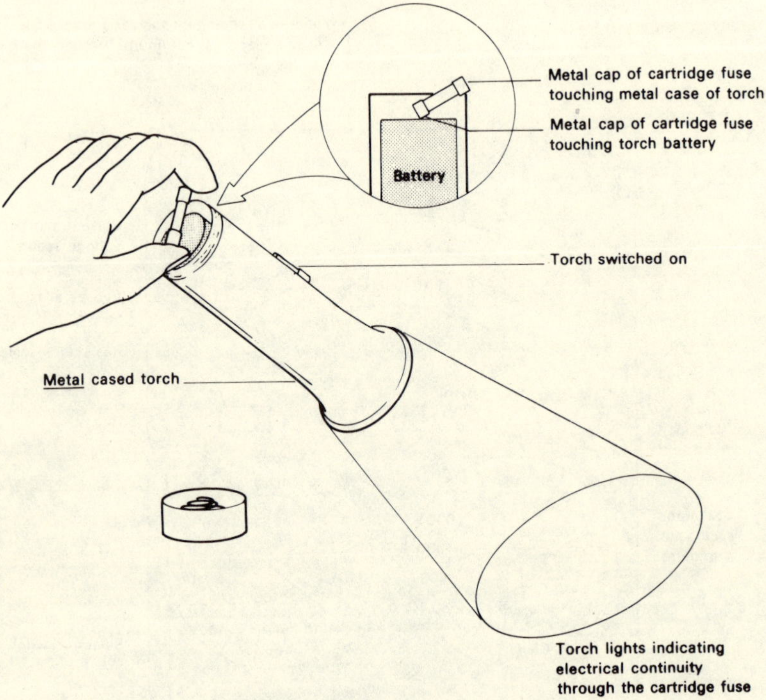

Fig.1.4 (b) Testing a cartridge fuse for continuity using a metal cased torch

A 13 A fuse would be suitable for this application.

The current rating of a fuse is based upon the maximum current in the circuit and not the current at which the fuse will operate. For example, in a 15 A circuit protected by a 15 A fuse the 0.50 mm copper wire in the fuse is designed to carry 15 A without overheating. It will not melt until a current of about 24 A is flowing in the circuit, i.e. it has a fusing factor of approx. 1.6. If the 15 A fuse provided melted as soon as 15 A (the nominal circuit current) were flowing the wire would become red hot and cause considerable damage to the distribution board or switchgear even before the circuit was fully loaded.

The correct fuses for appliances of different power ratings are listed in table 1.3. Cartridge fuses are colour coded for easy identification. Except for low power fuses where the wire can be seen through the glass envelope, it is not possible to identify a blown cartridge fuse from its appearance. If there is no obvious reason why an electrical appliance does not work after replacing a fuse then the fuse itself should be checked. This test may be carried out using a multimeter (see 9.10) to check for electrical continuity between the metal caps at opposite ends of the cartridge or by employing a simple torch battery and bulb circuit as shown in fig. 1.4. Equipment containing an electric motor and a number of other electrical appliances rated at less than 720 W may require a fuse of a higher rating than that calculated from its power consumption in normal operation as it may require a high starting current. The manufacturer's recommendations should be followed.

Table 1.3 *Recommended fuses and colour codes for 240 V supply*
(a) *Plug fuses*

Power rating of appliance	Recommended fuse /A	Colour coding
Up to 720 W	3	red
720 W to 2 kW	10	black
2 to 3 kW	13	brown

(b) *Main fuses*

Type of circuit	Recommended fuse /A	Colour coding
Lighting	5	white
Heating and general laboratory applications	15	blue
	20	yellow
Ring mains circuit	30	red and silver
Electric furnaces and other high power laboratory equipment	45	green
	60	purple

The commonest cause of overheating and breakdown of switchgear is incorrect fusing and not the overloading of the circuits they control. Incorrect fusing is responsible for more damage to electrical appliances and installations than any other single cause.

Cartridge fuses are preferable to rewirable fuses for the mains supply and for distribution boards. The rewirable fuse is considerably cheaper than the cartridge type and the cost of the short length of fuse wire required to repair a blown fuse is negligible. However, these fuses are subject to deterioration and are unreliable. There is also the possibility that the incorrect gauge of fuse wire may be fitted.

Assignment

Note the power ratings of the electrical appliances used at home and in the laboratories at work and in college. Use the relationship $I = W/V$ (see 1.4) to calculate the current drawn by the appliance and check that the appropriate size of fuse is fitted in the plug.

1.6
Selection of appropriately rated flex for a specified piece of equipment

The correct flex for a piece of electrical equipment is that which will allow the *safe* passage of the current drawn by the appliance as calculated in sub-section 1.4. It would be uneconomic for example to use an expensive heavy duty heating circuit cable rated at 30 A to connect a 60 W reading lamp to the mains. At the opposite extreme, it would suicidally unsafe to use a 3 A lighting circuit flex as the laboratory or workshop power supply cable. All flex and cable is rated according to the *maximum* current it will safely carry. This rating is given in the supplier's catalogue or is stamped on the spool or cable container along with a summary of possible applications, e.g. lighting, heating, heavy duty power machine circuits. Common, general purpose flex or cable is usually sheathed in p.v.c. for insulation and is available in a variety of ratings, e.g. 3 A, 6 A, 10 A, 13 A, 15 A and 18 A. It is dangerous to use a cable for a purpose in excess of its rated value (see 1.13).

The rating of flexible cord must not be less than the fuse protecting it. The current carrying capacity of a cable or flex may be defined as the maximum allowable current which may pass through it under the specified conditions without an undue rise in temperature or voltage drop.

(a) *Heating effect*

The energy loss by the heating effect of an electric current passing through a cable or flex depends on the resistance of the conductor and on the current which is passing:

$$\text{energy dissipated as heat} = I^2R.$$

(b) *Voltage drop*

The decrease in voltage which occurs when a current I passes through a cable of resistance R may be calculated by the use of Ohm's law:

$$\text{voltage drop} = I \times R.$$

A decrease in the voltage of a supply results in a loss of power and efficiency which may be apparent as poorer lighting, reduced heat from fires and decreased power from electric motors. A 10% drop in voltage produces a 20% power reduction. When connecting very long leads to an electrical appliance or supply it is often necessary to fit cable of a higher current rating to allow for the effects of voltage drop.

According to the IEE (Institution of Electrical Engineers) Wiring Regulation B23, the voltage drop between the supply terminals and any point in the installation must not exceed 2.5% of the declared voltage. The maximum voltage drop allowed for a supply from the 240 V mains is thus $2.5/100 \times 240 = 6$ V. It should also be remembered that the legally permitted variation in the declared voltage at the consumer terminals on an electrical distribution system is ±6% and there is a ±2.5% permitted variation in the frequency of the 50 Hz a.c. supply.

1.7
Conduction of electricity by the human body

The human body will conduct an electric current. Its electrical resistance varies widely from person to person and is strongly dependent on conditions, principally on whether the skin is dry or moist. If the skin is very dry its resistance may be as high as 10 000 ohms measured from hand to hand, nevertheless with a mains supply of 240 V this would allow a current of over 20 mA to pass through the body and could be fatal:

From Ohm's law,

$$I = \frac{V}{R}$$

where I = current;
V = voltage;
R = resistance;

therefore $$I = 240/10\,000 = 0.024 \text{ A}.$$

The resistance of wet or sweaty skin may be as low as 100 ohms. Under these conditions electricity can be very dangerous and severe injury or death can result with voltages considerably lower than that of the mains supply.

For this reason, switches, sockets and electrical equipment must not be placed close to taps, sinks and other areas where they may be splashed with water. Electric shock would be especially severe if a current passed through the wet skin of a person in a bath of water as the water would conduct the current to the pipework. The metal pipes are in direct electrical contact with earth so this route through the person's body would provide a low resistance path to earth and the electric shock which resulted would probably be fatal.

1.8
The physiological consequences of the passage of an electric current through the body

Very small currents (approx. 5 mA d.c. or 1 mA r.m.s. at 50 Hz a.c.)* passing through the skin produce a tingling sensation. Higher currents produce muscular contractions which at about 70 mA d.c. or 15 mA at 50 Hz a.c. (7 mA at 60 Hz) are so severe that the casualty may not be able to release his hold. If the current is increased beyond this 'threshold of muscular decontrol', there is considerable danger to life as irregular contractions of the heart result which quickly cause its normal pumping action to cease.

A current greater than 100 mA (a.c. or d.c.) would almost invariably be fatal if it passed right through the body, e.g. from one hand to the other or from a hand to a foot. Higher currents through a *part* of the body such as a hand or a finger may not be fatal, but can inflict severe burns in addition to electric shock. Even slight shocks can be dangerous as a sudden convulsion or movement can cause the victim to fall from a ladder or against a machine, for example, and suffer a more serious injury.

DO NOT TOUCH A PERSON SUFFERING FROM ELECTRIC SHOCK UNTIL YOU ARE CERTAIN THAT THE CURRENT HAS BEEN TURNED OFF (see 12.15).

1.9
Earthing electrical equipment

All electrical equipment not fully insulated must be properly earthed in the interests of safety. The correct connection of the green and yellow striped earth lead from the appliance to the centre terminal of the three-pin plug is generally all that is required (see 1.3). A qualified electrician should be consulted if the equipment is not already fitted with an earth lead. The end of the green and yellow striped cable should be attached to a terminal on the inside of the metal casing, preferably in the switch handle. When the plug is then placed into the earthed socket of the

---

* The commonly used abbreviations d.c. and a.c. refer to direct current and alternating current respectively. The hertz (symbol Hz) is the SI unit of frequency and is equal to the number of cycles of the waveform of the alternating current which occur in one second.

$$\text{frequency (s}^{-1}\text{ or Hz)} = \frac{1}{\text{time for 1 cycle}}$$

Common multiples of frequency include the kilohertz (kHz) and the megahertz (MHz), where

$$1\ \text{kHz} = 10^3 \text{ cycles per second}$$

and

$$1\ \text{MHz} = 10^6 \text{ cycles per second.}$$

An r.m.s. value refers to the root mean square value of an a.c. voltage or current.

power supply there will be a direct path to earth. A screw should always be used as well as solder when earthing metal-clad equipment. Simple soldered joints are not sufficient for a reliable earth connection as the soldered joint may be 'dry'.

1.10
Reason for earthing a metal clad electrical appliance

Some of the electrical equipment used in the home, laboratory or workshop does not require earthing as it is fully insulated, i.e. all the parts which carry an electric current are enclosed in a casing of plastic or other insulating material. In normal use it is impossible for the hand or other parts of the body to come into contact with the current and the appliance is thus completely safe unless the insulator is cracked or damaged. However, larger pieces of electrical equipment are frequently encased in a metal container as part of their structure or in order to mechanically protect (or 'sheath') fragile components. All metal clad appliances must be earthed so that should any fault—such as an incorrect or loose mains connection, for example—develop the chassis or metal casing does not become live. Anyone touching such faulty equipment would be in danger of electrocution as their body could provide a path of least electrical resistance to earth for the current and thus complete the electrical circuit (see fig. 1.5).

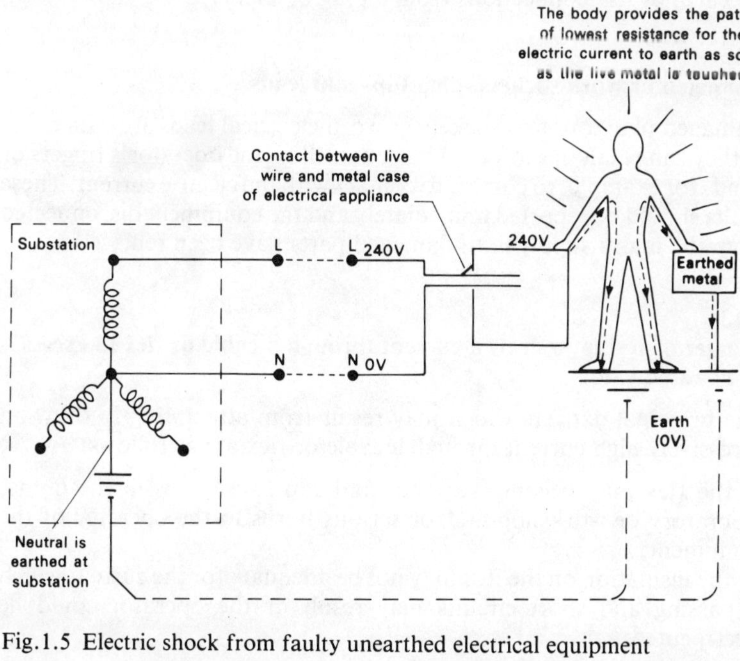

Fig. 1.5 Electric shock from faulty unearthed electrical equipment

Other methods of guarding against such dangers are:

1. the use of low voltage power supply to minimise the potential shock voltage;
2. the use of a double insulated construction which prevents electricity from returning to earth through the operator if an electrical fault develops in the equipment or wiring circuit; and
3. the fitting of an automatic earth leakage circuitbreaker which trips immediately any fault develops and cuts off the current flowing through the circuit.

1.11
Potential accidents in the laboratory resulting from the careless routing of equipment flex

When connecting electrical equipment to the mains, always use a length of cable just long enough to enable the connection to be made without straining it. Avoid trailing flex over the edges of benches or across the floor where it is susceptible to mechanical damage and where anyone tripping or catching the flex could bring electrical equipment crashing to the floor. Wherever possible cables should be firmly attached to the wall or bench and two or more leads should be strapped together into neat bundles and secured. Remember that unsecured lengths of cables can place an enormous strain on the terminals and could easily pull out the earth or live connections from a plug or socket.

1.12
Damaged or worn sockets, plug tops and leads

Damaged plug tops and sockets or worn electrical leads are dangerous as they can result in short circuits or can allow the operator's fingers or hand, for example, to come into contact with the electric current. These faults should be reported immediately and the equipment disconnected from the mains supply until damaged parts have been replaced.

1.13
Danger of trying to draw a current through a cable or flex in excess of its rated value

The principal dangers which may result from attempting to draw an excessively high current through a cable or flex are as follows:

1. the flex may become very hot and can cause fires in the home, laboratory or workshop and/or serious burns to the operator of the equipment;
2. the insulation on the flex may not be adequate for the current which is passing and short circuits may result or the operator could be electrocuted;

3. the heating effect of the excessively high current could melt the cable insulation, thus leaving the wires bare, and even cause it to break down chemically. This can be particularly dangerous with plastic insulating materials such as p.v.c. where the products of thermal degradation are highly toxic fumes.

1.14
Use of multiple adaptors and distribution boards

The current drawn by a number of electrical appliances connected in parallel is additive. If this current is drawn from a single supply via a multiple adaptor or distribution board the operator must always check that the *total* current drawn does not exceed the maximum safe value

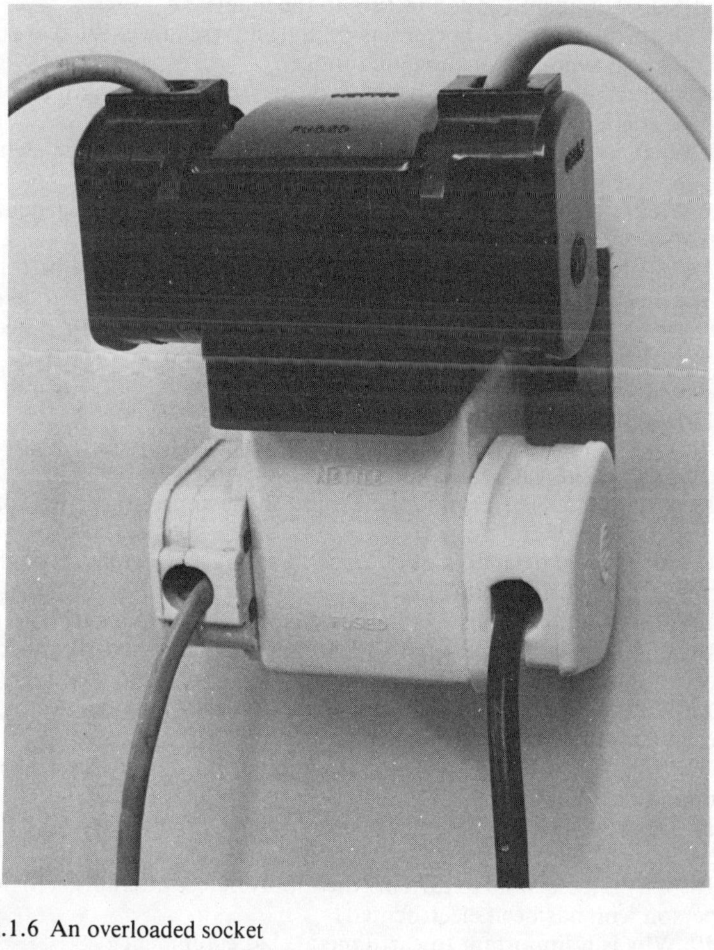

Fig.1.6 An overloaded socket

for the circuit or the hazards described in sub-section 1.13 may result (see fig. 1.6).

The use of a large number of plugs and adaptors in a single socket should be avoided as it can:

1. break the socket;
2. pull the wiring and socket away from the wall;
3. create intermittent contact which causes arcing and superheating and can lead to fires.

Questions: 1 *Hazards of electrical equipment*

1.1 What is the BS colour coding for cable and flex?
1.2 Discuss the implications of colour blindness to the wiring of electrical equipment for connection to the mains.
1.3 Distinguish between current, potential and resistance. What are the SI units and symbols for these quantities?
1.4 What is the relationship between the resistance of a length of wire and its conductivity?
1.5 What is meant by the terms 'conductor' and 'insulator' when applied to electricity? Give *three* examples of each.
1.6 What is meant by the 'power rating' of a piece of electrical equipment?
1.7 Calculate the current drawn by the following appliances operating on a mains voltage of 240 V: (a) a 5 kW heater, and (b) a 750 W heater.
1.8 What determines the choice of the appropriate fuse for a given application?
1.9 Why is it dangerous to replace a blown fuse with a short length of copper wire?
1.10 Why is it unwise to jam the switch of an earth leakage trip unit in the ON-position if it continually cuts off the supply?
1.11 What factors determine the choice of the appropriate flex for a specified piece of equipment?
1.12 What are the dangers of attempting to draw a current through a flex in excess of its rated value?
1.13 Why is it dangerous to fit a power socket close to a sink?
1.14 Why is it inadvisable to fit a 13 A fuse in the plug used to connect a piece of laboratory equipment rated at 250 W to a 240 V supply?
1.15 What action should be taken if the flex and plug connecting a piece of equipment to the mains becomes hot?
1.16 What are the factors which determine the conductivity of the human skin?
1.17 Describe the physiological consequences of the passage of an electric current through the human body.
1.18 Why is it essential to turn off the supply before attempting to help a person who has been electrocuted?
1.19 Why is it important to earth metal clad equipment?

1.20 Describe how you would earth a piece of electrical equipment.
1.21 Electricity can cause injuries other than burns and shock. Describe how these secondary injuries can occur.
1.22 Why is it important to replace a mains lead if the insulator is cracked or hardened?
1.23 Why is it unsafe to route power cables across the floor?
1.24 Why is it important to know the exact location of the mains electricity switch in the laboratory?
1.25 Explain why electrical faults are a common cause of fires.
1.26 State whether it would be safe to operate appliances with the following power ratings from a multiple socket outlet panel (fig. 1.7) connected to the mains by a cable rated at 10 A: (a) 50 W, 1 kW, 500 W, 250 W and 100 W; (b) 1 kW, 2 kW, 0.5 kW, 100 W and 750 W. In each case suggest an appropriate fuse for the plug.

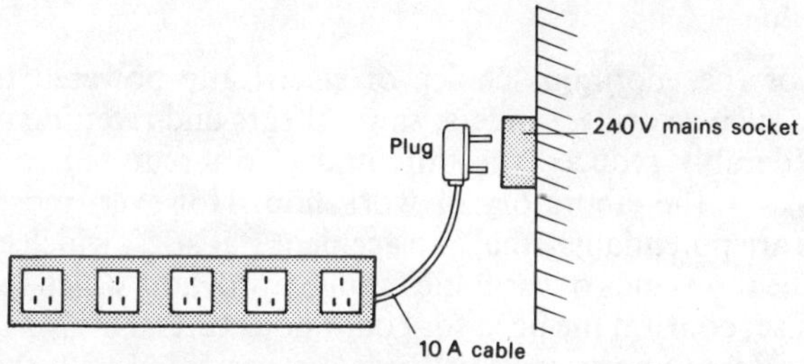

Fig.1.7 A multiple socket outlet panel

Section 2: *The expected learning outcome of this section is that the student should be able to state the elementary safety precautions which should be borne in mind when using power tools*

Specific objectives: *The expected learning outcome is that the student*:

2.1 *Wears appropriate protective clothing if required.*
2.2 *Recognises the hazardous nature of loose clothing such as ties.*
2.3 *Recognises the hazards associated with long hair and any power tool.*

Introduction

The use of the enormous choice of electrically powered tools now available, such as drills, sanders, saws, shears and rivetting machines, has considerably reduced the time and effort required to complete many tasks in the laboratory or workshop. However, most of these machines are potentially lethal and accident statistics, which every year disclose the thousands of casualties and many deaths which result from their misuse, confirm the need for continuous care. This is particularly true of portable power tools as these are not fitted with the guards, screens, barriers, automatic cutoff switches and other protective devices of the fixed machines.

Most accidents involving power tools are caused by incorrect handling or inadequate maintenance. Carelessness, ignorance, inadequate training, over-familiarity, horseplay, lack of attention, inadequate insulation and frayed electrical leads have all produced their share of injury and death. For the safe operation of power tools all these hazards must be removed. The equipment must be correctly earthed or double insulated (see 1.10). The international sign for double insulation and the British Standards Institution Kitemark for portable electric tools are shown in fig. 2.1. These signs provide an independent guarantee of safety to the operator (see 9.16). About a half of all the fatal electrical accidents in industry involve the use of portable equipment. No accidents have been reported involving electric shock to the operator while using power tools of double insulated construction. The manufacture of electric hand tools which operate at lower voltages than that of the mains supply is an additional safety measure. The manufacturer's recommended operating procedure should always be followed and the elementary safety precautions concerning the wearing of protective clothing and the hazards associated with long hair or loose clothing such as ties or scarves must be observed.

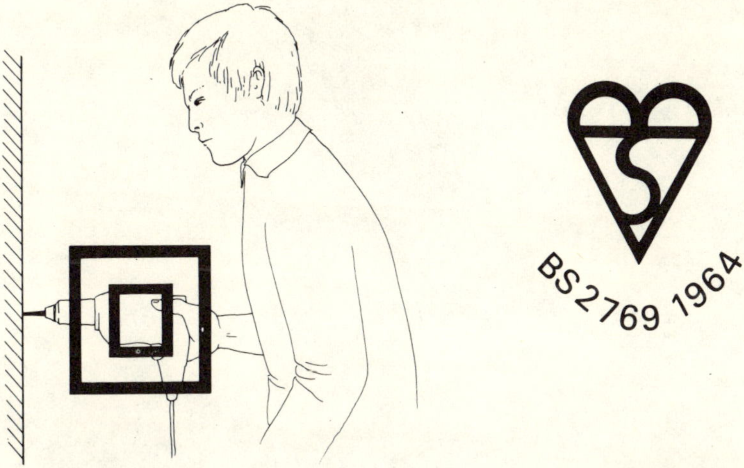

Fig.2.1 The international sign for double insulation, and the BSI Kitemark

2.1
Appropriate protective clothing

A boiler suit is perhaps the most appropriate protective clothing to wear for the *regular* operation of power tools; however, for occasional use a well fitting linen laboratory coat is equally suitable. A face shield or goggles (fig. 2.2.) should be worn to protect the eyes from grit or flying splinters of metal and a dust mask (fig. 7.2) is recommended for sanding, grinding and other operations which produce large quantities of fine dust. Gloves or water soluble, mildly antiseptic barrier creams which may be removed by washing protect the hands while handling greasy, hot, rough or oily materials.

2.2
Hazards of loose clothing

All loose clothing is potentially hazardous when using power tools. Ties are particularly dangerous as they easily wrap round any piece of revolving machinery and drag the operator towards it. Scarves, the torn ends of overalls, necklaces, keychains, loose cuffs or unbuttoned lab coats can also be caught up in the moving parts and should not be worn.

2.3
The hazards associated with long hair and any power tool

Long hair is a serious hazard when operating any power tool or machinery with moving parts. The hair can easily become entangled

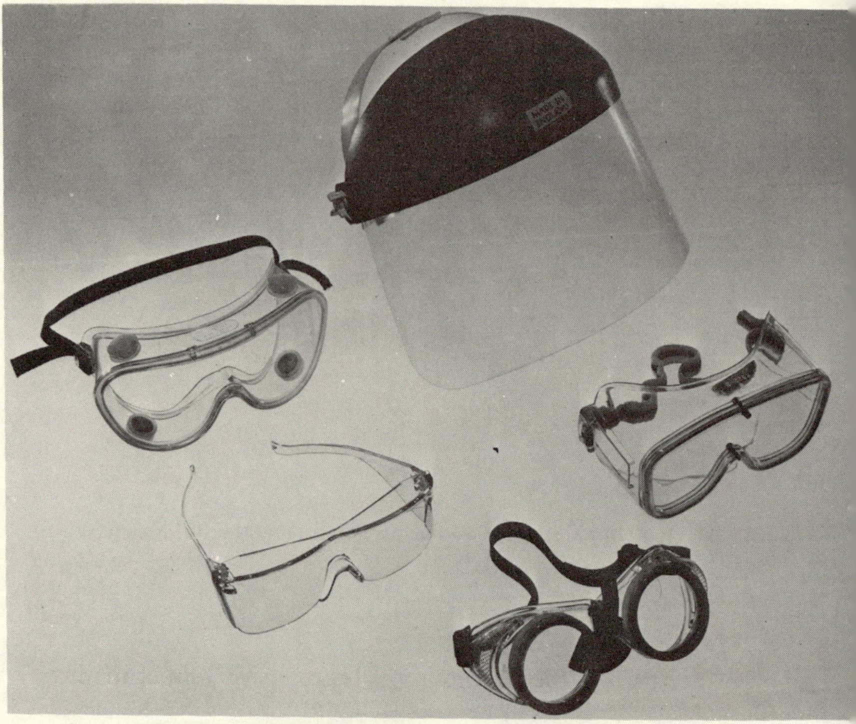

Fig.2.2 Face shield and safety goggles

and scalp the operator or, at least, the hair could be torn out by the roots. The result in either case is extremely painful and can cause death or permanent brain damage. Long hair must be tied back and covered with a suitable cap before using any power tool. The hair covering may not be particularly fashionable, but at least it is preferable to having one's scalp ripped off the skull.

Assignment

What protective clothing is available for an operator using power tools in your laboratory at work? Do you consider this adequate in view of the hazards involved? What precautions are taken when using an electric stirrer in the laboratory?

Questions: 2 *Using power tools*

2.1 Why are portable power tools the cause of so many accidents in industry?
2.2 What are the hazards of using a power tool if the operator has long hair or is wearing a necklace or tie?

Section 3: *The expected learning outcome of this section is that the student should be able to recognise the elementary hazards which may be encountered in a chemical laboratory*

Specific objectives: The expected learning outcome is that the student:

3.1 *Uses a fume cupboard.*

3.2 *States that poisons should be kept locked away and logged in and out of storage.*

3.3 *Uses supplied reference material to determine whether a given chemical is toxic, inflammable or otherwise dangerous.*

3.4 *States that inflammable liquids should be stored, in bulk, away from main buildings.*

3.5 *Safely transports from store bulk chemical containers.*

3.6 *Recognises the need for always labelling chemical containers.*

3.7 *States the necessity for and reasons for scrupulous attention to hygiene.*

3.8 *States the nature of the hazards associated with damaged glassware, cutting tools etc.*

3.9 *Lists the common carcinogenic chemicals.*

3.10 *States the precautions to be employed if carcinogens do have to be handled.*

Introduction

The numerous hazards of the chemical laboratory are mainly a result of the many different harmful properties which chemicals can possess. Thus these substances may be poisonous, inflammable, corrosive or caustic; they can affect the nerve cells or irritate the skin and cause dermatitis and skin allergies. Some are carcinogenic, i.e. they are capable of producing cancer, while others (such as ether or trichloromethane (chloroform)) act as anaesthetics and are occasionally addictive. Other hazards come from cutting tools, broken glassware, hot apparatus, burning gas and the use of electrical equipment, gas cylinders, compressed air and vacuum systems.

All of these factors represent very real, potential dangers, but provided these are recognised and the appropriate action is taken there is no reason why the average chemical laboratory should be less safe to work in than any other place of employment. The accident statistics support this and far fewer injuries are reported in laboratories than occur on sports fields, stairs, crossing roads or in the home.

Most accidents in a chemical laboratory are a result of ignorance or carelessness. It follows therefore that these accidents will be avoided with a proper awareness of the nature and dangers of the materials involved and by maintaining a strict code of behaviour (including the use of appropriate protective clothing) at all times in a laboratory. The use of reference literature to determine the chemical hazards of substances is discussed in sub-section 3.3 and a safety code for anyone working in a laboratory is described in sub-section 7.2. The main hazards are discussed more fully in this chapter. For example, a jet of compressed air should never be directed against the skin, especially if it is scratched or cut as this can force air into the bloodstream and cause injury or death. Compressed air should not be used for clearing dirt or swarf from crevices behind machines etc. as the flying fragments will be propelled with considerable force.

Safety screens and wire-mesh guards should be placed round evacuated glassware (see fig. 3.1) and safety goggles or a face shield must be worn against the possibility of implosion. This particularly applies when large desiccators or Buchner funnels are being evacuated. Thin-walled glassware must never be connected to a vacuum line or filter pump. Glassware should be examined for flaws before it is evacuated. It should not be heated as the difference in thermal expansion of the inner and outer walls of a thick-walled flask, for example, will set up stresses in the glass which could cause it to crack. Atmospheric pressure (101 325 Nm^{-2}) is equivalent to a force of more than 1 kg weight on every square centimetre (or 14 lb in^{-2}) of the glass surface or a total of 0.5 tonne spread over the whole surface of a 1 dm^3 flask. If this crack occurred when the vessel was partially evacuated the

Fig.3.1 A Buchner funnel and vacuum desiccator with protective wire-mesh guard

effect of this enormous force compressing the weakened flask would cause it to implode (i.e. to '*ex*plode inwards') propelling fragments of broken glass in all directions.

The dangers associated with cylinders of compressed gases should also be recognised. These hazards come from three sources, *namely* the contents of the cylinder, its weight and the high pressures contained within it. A cylinder should be painted in the appropriate colours of the British Standards Specification 349 : 1973 (see table 3.1) to identify its contents, which should also be written in words on the side. The colour is intended only as a secondary guide. The *written name* is the only legally recognised method of identifying the cylinder contents. Care should be taken with cylinders from American companies as the USA colour code is different from that used in Britain.

Table 3.1 *Identification colours for gas cylinders*

Gas	Colour of cylinder	Colour of bands* (if any)
Air	Grey	—
Ammonia	Black	Red and golden yellow
Argon	Blue	—
Carbon dioxide	Black	—
Carbon monoxide	Red	Golden yellow
Chlorine	Yellow	—
Ethyne (acetylene)	Maroon	—
Hydrogen	Red	
Nitrogen	Grey	Black
Oxygen	Black	—
Sulphur dioxide	Green	Yellow

* The colour of the band(s) on the shoulder of the cylinder indicates the particular hazards associated with the contents:

a *red band* indicates a flammable gas;
a *yellow band* indicates a toxic gas;
a *red* band *above* a *yellow* band indicates a flammable, toxic gas.

The contents of compressed gas cylinders can range from highly inflammable substances, such as hydrogen, propane and ethyne (acetylene), to toxic gases, such as chlorine or carbon monoxide, and to comparatively inert substances like argon and nitrogen. Cylinders of toxic gases should not be left in the laboratory overnight in case of leaks. Large cylinders may weigh as much as 85 kg and should always be clamped vertically (fig. 3.2) to prevent them toppling over or laid horizontally on the floor and wedged to prevent rolling. Damage could

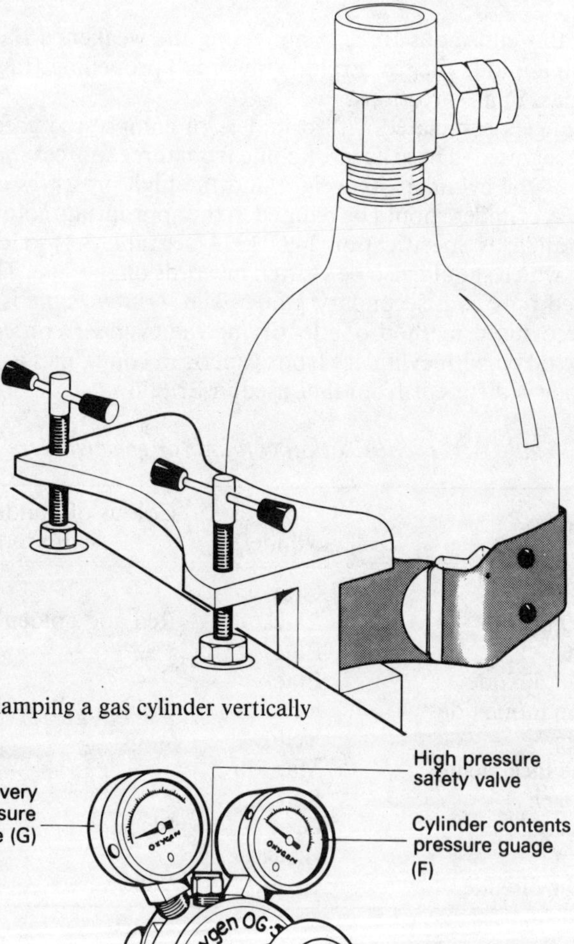

Fig.3.2 Clamping a gas cylinder vertically

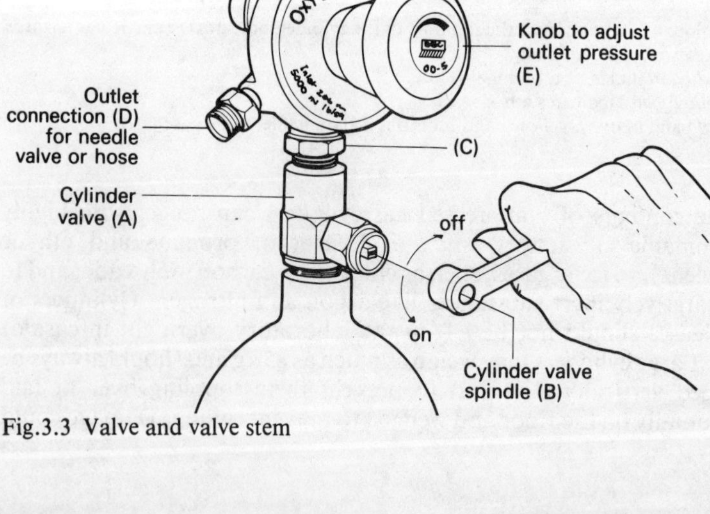

Fig.3.3 Valve and valve stem

cause the base of the valve stem (fig. 3.3) to shear off and sudden release of pressure can propel a heavy gas cylinder across the room or through brick walls like a shell. Compressed gas cylinders frequently contain pressures in excess of 2500 lbf in^{-2} or 180 atm (18 MN m^{-2}), i.e. more than 1 ton per square inch. They must not be heated and care must be taken when moving cylinders. The use of a cylinder trolley (fig. 3.4) is recommended. Stiff cylinder valves must be opened carefully and not forced with hammers or wrenches with excessive leverage. Oil or grease must never be used on lines, reducing valves or other equipment for oxygen cylinders as this can cause explosions. Centrifuges are also a potential source of danger. The outer edge may be moving at speeds in excess of 150 km h^{-1} so it should not be touched when it is operating, nor should the machine be stopped with the hands. A fully enclosed construction in which the motor cuts out as soon as the lid is lifted should be employed wherever possible. Cracked or damaged centrifuge tubes must never be used. A tube should always be balanced with another tube filled with liquid to the same depth to ensure even running.

3.1
Fume cupboards

Any reaction or process which produces harmful gases or vapour must be carried out in a fume cupboard and not on a bench in the open laboratory. Benzene, fuming acids, phenylamine (aniline) and other liquids which produce a toxic, irritating, flammable or otherwise

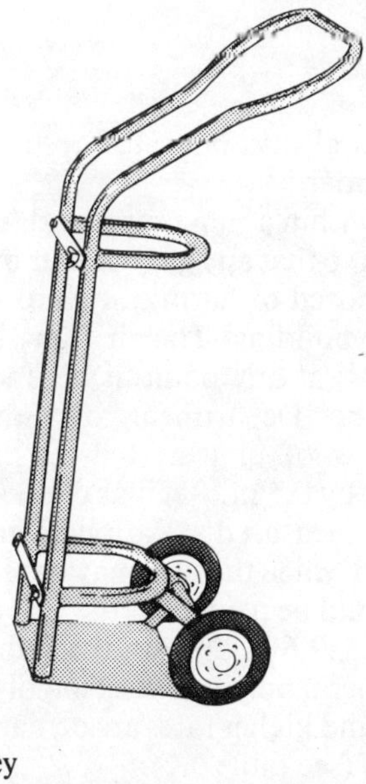

Fig.3.4 A cylinder trolley

Fig.3.5 A fume cupboard

harmful vapour should always be transferred from one container to another in a fume cupboard.

The enclosed space within a fume cupboard (see fig. 3.5) is linked to an air extraction system so that any gas, vapour or harmful dust is swept out into ducts and disposed of harmlessly into the atmosphere via an outlet at the top of the building. The air flow should not be less than 0.75 m s^{-1} (150 ft min^{-1}) at any point at the face of the cabinet. For school laboratories the Department of Education and Science recommend an air flow of at least 0.3–0.5 m s^{-1} when the fume cupboard sash is raised by 0.5 m. Regular checks should be carried out and the flow should be measured at various points along the length of the frame. The points to which the sash may be raised to give the 0.3 and 0.5 m s^{-1} flow rates should be painted at the side of the fume cupboard. An entry velocity of 0.5–0.6 m s^{-1} with the fume cupboard sash fully open is required for general purpose work in polytechnics, universities, hospitals and industry and higher rates are demanded for prolonged use of hazardous materials (see table 3.2).

Table 3.2 Recommended entry velocities through fully open sash for prolonged use of specially hazardous substances

Material	Recommended entry velocity /ms^{-1}
Perchloric acid, hydrofluoric acid	0.500–0.750
Organic solvents	0.500–0.750
Bromine, chlorine, cyanides and other highly toxic chemicals	0.500–0.750
Radioactive substances	0.500–1.025

The sliding panels which provide access to the fume cupboard should be fitted with safety glass and not with ordinary window glass. Laminated and (preferably) unbreakable glass are particularly recommended. Potentially explosive reactions can then be carried out in an enclosed space which will safely contain the fumes, fragments of flying glass and other products of the explosion without risk to the spectators.

Fume cupboards should be cleared out regularly. With few exceptions—such as a Kipp's apparatus to generate hydrogen sulphide—they should not be used to store chemicals or equipment. The practice of storing fuming liquids or toxic substances in fume cupboards produces a serious additional hazard in laboratories. The labels on the containers quickly become illegible and disintegrate to leave a stock of dusty bottles of which only one thing is known: the unidentified contents are dangerous.

Assignment

Examine the fume cupboards in the college laboratory and at work. Is there any indication of the height to which the fume cupboard may be opened before the through draught flow rate becomes inadequate?

3.2
Storage of poisons

Toxicity is a relative term as virtually all chemicals are harmful to the human body if taken in excessive amounts. With some substances there is a borderline between relatively safe and hazardous levels of exposure. The *threshold limit values* (t.l.v.) of toxic gases, vapours and dusts represent the weighted average concentration to which workers may be repeatedly exposed to the substance without adverse effects.

Two different units are in common use for expressing the concentrations of such extremely dilute solutions or mixtures with air:

1. mg/m^3, i.e. the mass of solute in 1 cubic metre of solution or air.
2. parts per million (which is abbreviated to p.p.m.)—this is defined as the number of parts of solute in one million parts of the mixture. The word 'part' may refer to any unit (e.g. grams, kilograms, tonnes or litres) as the term p.p.m. is a ratio and whatever units are used to express the amounts of solute and solution cancel out. A concentration of 1 p.p.m. is very small. The ratio is the same as that represented by one minute in a little less than two years or 1p in £10 000. For aqueous solutions, 1 p.p.m. = 1 mg per litre. Nevertheless, this does not mean that the dangers resulting from such small concentrations of toxic substances can be ignored. With some materials, it may be that any exposure—however slight—may be dangerous and significant amounts of a substance present in the air at a 5 p.p.m. level would be breathed in (and out) in the course of a working day.

The threshold limit values of a number of common laboratory chemicals are listed in table 3.3 as a guide for the control of hazardous materials and not as an indication of the conditions which should be considered acceptable by people working in laboratories.

Table 3.3 *Threshold limit values of common substances*

Substance	Threshold limit values (t.l.v.) $/mg\ m^{-3}$	/ppm
Ammonia	18	25
Benzene	80 (skin: 32)	(10)
Bromine	0.7	0.1
Chlorine	3	1
Hydrogen chloride	7	5
Iodine	1	0.1
Mercury	0.05	
Sulphur dioxide	13	5
Tetrachloromethene (carbon tetrachloride)	65	10
Trichloromethane (chloroform)	120	25
Trioxygen (ozone)	0.2	0.1

The very low threshold limit value of mercury is particularly significant in view of the wide use of this substance in all types of laboratory. The t.l.v. is less than 1% of the mercury vapour concentration equivalent to its saturated vapour pressure at room temperature. The vapour from less than 0.02 g of mercury is sufficient to contaminate all the air in the average sized laboratory to a concentration higher than that of the t.l.v. All mercury surfaces should be kept covered and any spilt mercury should be cleaned up at once. Any drops or droplets lodged in cracks or crevices from which it cannot be removed should be sprinkled with sulphur. Mercury or mercury oxide must never be heated in the open laboratory.

Some substances, such as benzene and lead, are cumulative poisons. Particular care must be taken in the use of these materials as the effects of frequent or irregular exposure to even small amounts of the substance may not be apparent for many months or years.

The potential health hazard of a substance is determined by its physical properties and its chemical structure. Solubility and volatility are important factors in this respect and fat-soluble substances, such as benzene or phenylamine (aniline), are also dangerous by skin absorption (see also 12.10). Most organic solvents will dissolve the protective secretions of the skin and continued exposure to some of these compounds can produce allergic reaction, dermatitis or even cancer.

The chemicals used for laboratory experiments are not subject to the provisions of the Poisons Acts; nevertheless, it is essential that all dangerous substances, such as cyanides, are always kept under close control. All scheduled poisons and substances of high toxicity must be kept in a locked cupboard or store and logged in and out as required. The key should be accessible only to responsible workers in the laboratory and the contents of the poison store should be checked at regular intervals. Only the quantities required for specific experimental purposes should be issued and then in clearly labelled containers (see 3.6 and 6.3). No experiment involving poisons should be carried out without knowing what action to take in case of accident and the required antidotes should always be at hand.

Assignment

Where are scheduled poisons stored at college and at work? Are these substances logged in and out when they are used?

3.3
Use of reference material to determine chemical hazards

Before any efficient action can be taken to minimise the hazards associated with using a particular substance it is necessary to know what the precise dangers are. The label on the container provides a guide (see 3.6. and 6.3.) and indicates, for example, whether the substance is inflammable, toxic or corrosive. This information may be sufficient if only small amounts of the material are being used on a single occasion. However, before carrying out extensive experimental work or purifying the material, reference should be made to published sources to obtain full details of the substance's physical and chemical properties, of its particular hazards and of the disposal procedure or first aid treatment to adopt in the event of spillage or an accident. The hazards associated with a number of common chemicals are listed in table 3.4. Comprehensive information on a wide range of hazardous chemicals has been published (see p. 209) and reference to this material is recommended.

Table 3.4 *Hazards of common laboratory chemicals and solvents*

Name of compound	Boiling point	Hazardous properties
Benzene	80 °C	Extremely flammable. Vapour very poisonous with danger of cumulative effects. Poisons by skin absorption. Benzene causes kidney and liver damage and it can destroy bone marrow and thus cause severe anaemia and other blood disorders which may be malignant or fatal. Carcinogens are often present as impurities in benzene. It has been recommended that the use of benzene as a solvent should be banned.
Chlorine	– 34 °C	Toxic gas. Irritating to the eyes, skin and respiratory system. It can cause conjunctivitis and severe lung damage.
Formalin (an aqueous solution containing approx. 40% of methanal (formaldehyde) and 11–14% of methanol)	96 °C	Flammable. Poisonous by inhalation and by swallowing. It causes burns and severely irritates the eyes, skin and respiratory system. Can cause ulceration and cracking of the skin.
Propanone (acetone)	56 °C	Highly flammable, forms explosive mixtures with air (see table 5.2). The vapour irritates the eyes and can produce cataracts. Inhalation may cause dizziness and coma.
Tetrachloromethane (carbon tetrachloride)	77 °C	Poisonous vapour. Inhalation causes headaches, mental confusion, depression, fatigue, nausea, vomiting and coma.

Table 3.4 *(Continued)*

Name of compound	Boiling point	Hazardous properties
		Repeated contact with the liquid can produce dermatitis. Taken by mouth it can damage the liver, kidneys, heart and nervous system. Even small doses may be fatal. It can induce cancer.
Methanol	65 °C	Highly flammable. Poisonous by inhalation, swallowing and skin absorption. Inhalation causes dizziness, nausea, cramps, headache and vomiting. If swallowed, the liquid can damage the central nervous system and cause blindness (some methanol and unpleasant smelling substances such as pyridine are added to ethanol in methylated spirits to *denature* the alcohol and make it unfit for drinking). Methanol can also injure the kidneys, liver and heart. Unconsciousness may develop some hours after taking methanol and death may follow.

These references also provide information about hazards which may develop in a material on standing or when the substance is treated with a particular chemical. Ethoxyethane (ether), for example, slowly forms explosive peroxides on standing in the presence of light and air. These peroxides have a higher boiling point than the ether and therefore form a dangerous residue during distillation. The presence of peroxides may be detected by the liberation of iodine when a sample of the ether is added to aqueous potassium iodide solution. Peroxides may be destroyed by shaking the ether with iron (II) sulphate solution. The distilled material is then stored in dark glass bottles containing a small coil of clean copper wire. The bottles should be of such a size as to be almost filled with the liquid as this limits the amount of air and oxygen in contact with the ether. A number of hazardous mixtures or reactions are listed in table 3.5 for reference.

Table 3.5 *Hazardous mixtures and reactions*

Substance	Violent reaction when mixed with:
Alkali metals (e.g. sodium or potassium)	Water, acids, carbon dioxide, tetrachloromethane and other chlorinated hydrocarbons.
Aluminium (especially in powdered form)	Chlorates, nitrates.
Ammonia	Chlorine, bromine or iodine. Calcium hypochlorite. Mercury.
Ammonium nitrate	Metal powders, sulphur, chlorates, powdered organic compounds.
Chlorates	Metal powders, sulphur, ammonium salts, finely divided organic compounds or other combustible materials, sulphuric (VI) acid, picrates.
Chlorine	Ammonia, ethyne (acetylene), hydrogen, finely divided metals.
Chromic acid	Ethanoic (acetic) acid, glycerol, naphthalene, alcohol and other flammable liquids.
Copper	Ethyne (acetylene), hydrogen peroxide.
Ethanoic (acetic) acid	Chromic acid, nitric acid, glycol and other hydroxy compounds, manganates (VII), peroxides and perchloric acid.
Ethyne (acetylene)	Chlorine, bromine and iodine. Copper, silver or mercury.
Hydrocarbons	Chlorine, bromine, chromic acids, peroxides.
Hydrogen peroxide	Copper, chromium, iron and many other metals or their salts. Any flammable liquid or combustible material.
Hydrogen sulphide	Fuming nitric (V) acid.
Iodine	Ammonia (gaseous or aqueous solution), ethyne.
Mercury	Ethyne, ammonia.

Table 3.5 *(Continued)*

Substance	Violent reaction when mixed with:
Nitric (V) acid (conc.)	Ethanoic acid, alcohols, propanone (acetone), phenylamine (aniline), hydrogen sulphide, flammable liquids or solids, chromic acid.
Oxygen	Grease, oil, hydrogen and any other highly flammable materials.
Potassium manganate (VII) (permanganate)	Glycol, glycerol, sulphuric (VI) acid, benzaldehyde.
Propanone (acetone)	Mixtures of concentrated nitric (V) and sulphuric (VI) acids.
Sodium nitrate (III) (nitrite)	Ammonium nitrate (V).
Sulphuric (VI) acid	Chlorates, perchlorates, permanganates. Water (dilute acid by adding acid to water and not *vice versa*).

Assignment

List the six chemicals you use most frequently at work and with the help of suitable reference material outline the potential hazards (if any) involved in their use.

3.4
Storage of inflammable liquids

Bulk supplies of all flammable liquids should be kept in a solvent store well away from the main buildings. The store should be securely locked and fire warning notices (see 6.3) should be prominently displayed on the door. Electric switches for lights should be of a 'spark free' construction to prevent ignition of spilled solvent vapour. For the same reason, safety lights in which the hot surface of the electric light bulb is contained within a glass cover should also be fitted (see sub-section 5.1(c)). This isolation of large drums and Winchester bottles of flammable solvents etc. considerably reduces the extent and likelihood of laboratory fires.

Bottles of liquids must not be placed in direct sunlight. The liquid contained within the curved glass can act as a lens to focus the sunlight. Considerable temperature increases can be obtained which may result in a fire. A steel bottle store (fig. 3.6) is suitable for keeping the small amounts of inflammable liquids used in many schools and other small

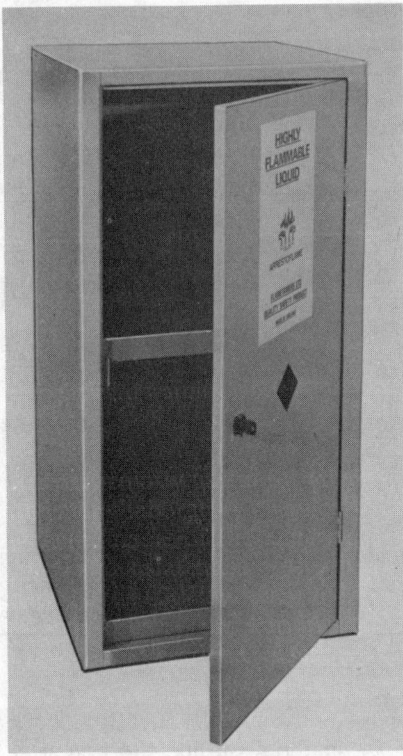

Fig.3.6 A steel store for bottles of flammable liquids

laboratories. Naturally these containers should be properly labelled to indicate the presence of a fire hazard and should not be placed anywhere near radiators or any naked flames.

Assignment

What provision has been made for the storage of bulk supplies of inflammable liquids in your laboratory at work?

3.5
Transport of bulk chemicals from store

Ideally only sufficient concentrated acids, flammable solvents and other hazardous chemicals for immediate requirements should be kept in the laboratory. Larger amounts, e.g. 500 cm^3, 1 litre and 2½ litre (Winchester) bottles, carboys and metal drums, should always be kept in a separate store (see 3.4). Acid-resistant trays or troughs are recom-

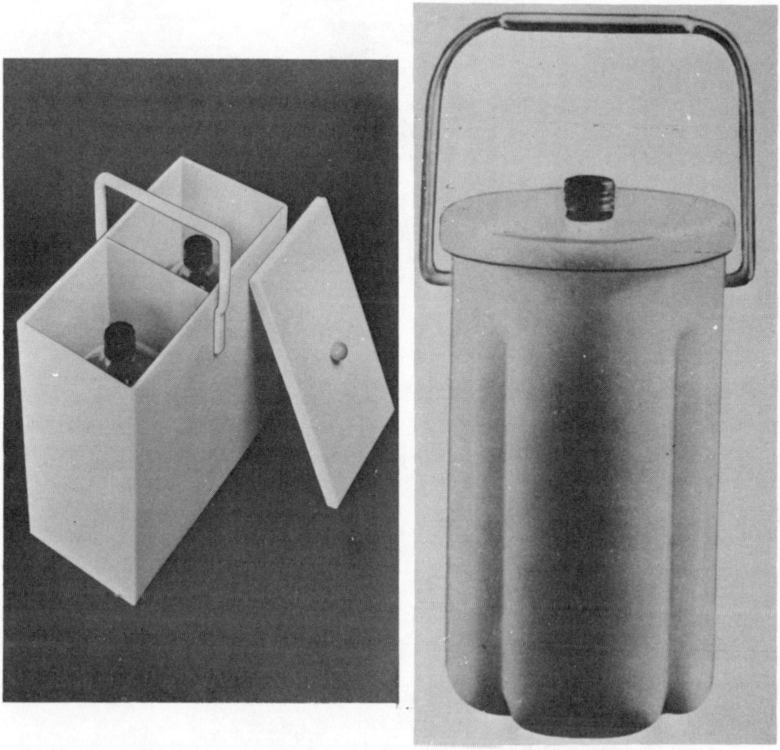

Fig.3.7 Carriers for Winchester bottles

mended for the storage of concentrated sulphuric (VI), nitric (V) and other acids. They are also useful for storing all liquids as the tray will contain the contents of any bottle which breaks and thus prevent harmful liquids soaking into shelves and spilling over onto other containers or onto the floor.

Winchester bottles should never be carried by the neck as the bottle can easily slip out of the hand and smash on the floor, while occasionally the thickness of the glass is not sufficient to bear the weight of the contents. Nor should these bottles be carried in the arms or in the hands. A proper carrier, such as a wickerwork basket or the polythene container shown in fig. 3.7, must always be used for transporting Winchester bottles from the store or from one laboratory to another.

Carboys of concentrated acids and other liquids should be vented or the increase in pressure as the contents heat up on being brought from a cold store into a warm room is sufficient to burst the carboy. Whenever possible carboys should be left in the bulk store and liquids dispensed as required using a carboy siphon (fig. 3.8).

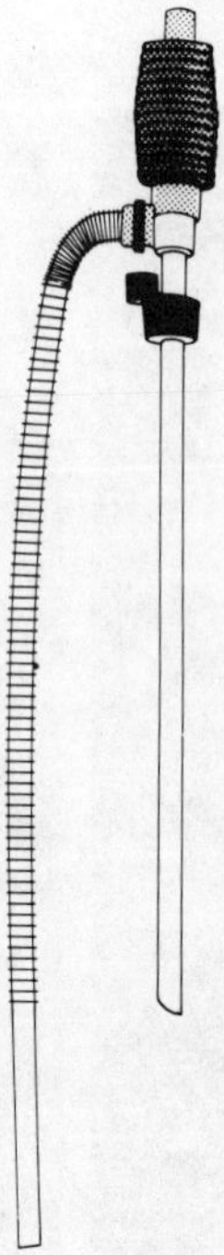

Fig.3.8 A carboy syphon

Assignment

What equipment is available in your laboratory for transporting Winchester bottles from the bulk store or for dispensing liquids from carboys and metal drums?

3.6
The need for always labelling chemical containers

The reason for always labelling chemical containers is obvious. The technician working in the stores or preparation room, for example, may know that the substance he has transferred to an unlabelled container is poisonous and is explosively unstable on heating or that the liquid filling an unlabelled Winchester is freshly prepared battery acid or formalin, but unless others are aware of the fact the dangers from mistaken use may be serious. At busy times of the day when perhaps a number of solutions or reagents are being prepared for different experiments the probability of an accident occurring increases. All these accidents are avoidable in a well organised laboratory.

It should be remembered too that the chemicals stored on laboratory shelves may be used by people whose knowledge of chemistry is insufficient to make them aware of the dangerous properties of the materials concerned. All chemical containers should be clearly labelled to show not just the name, but also the potential hazards of the contents. Labels should be written with a ballpoint pen or preferably with indelible ink so that the words do not fade or run on contact with water or organic solvents and then firmly attached to the container. Gummed labels are not sufficient as they easily peel off when dry or if they become damp. They should therefore be covered with a strip of 'Sellotape'. The name of the contents should never be written directly on the glass with a felt-tipped pen as the writing is quickly obliterated as soon as someone handles the bottle. Beakers or conical flasks of solutions for use in a titration, for example, may be distinguished by writing the name or chemical formula of the contents on the small white disk on the side of the container in pencil. Naturally, to avoid future confusion, the symbols should be erased when the glassware is cleaned at the end of the experiment.

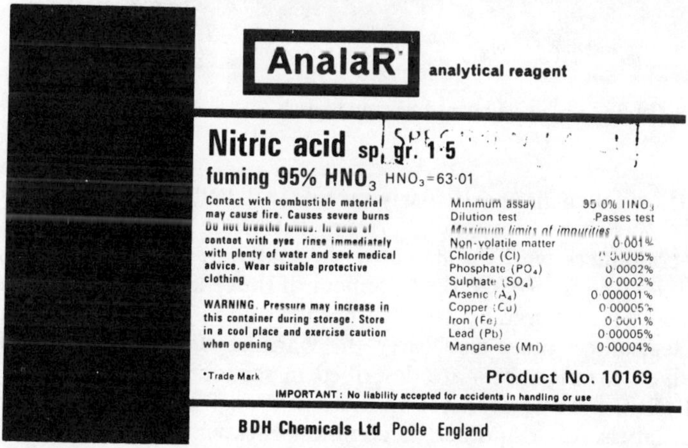

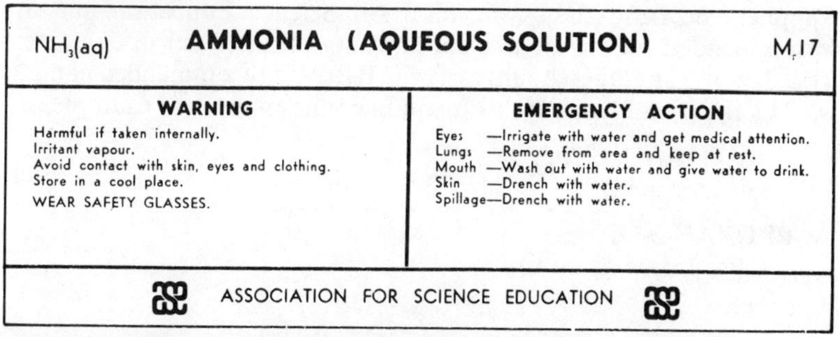

Fig.3.9 (a) BDH and ASE labels (M_r = Relative molecular mass)

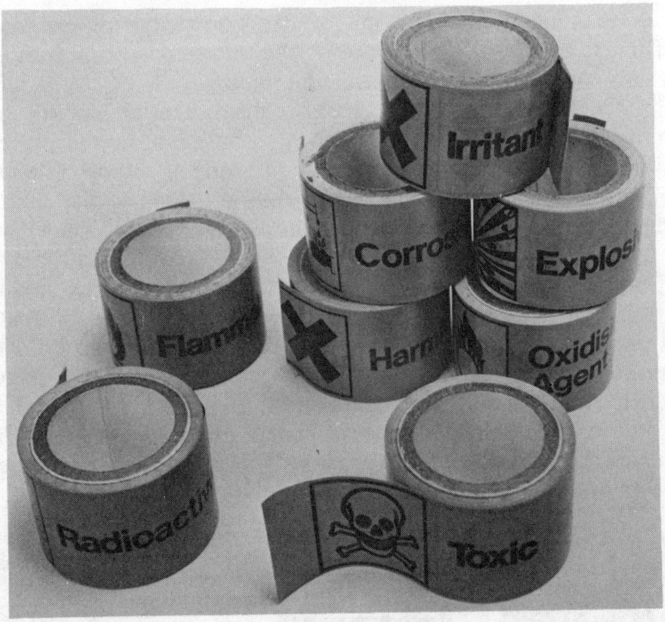

Fig.3.9 (b) Self adhesive hazard warning labels

Hazard labels are particularly useful for labelling jars and reagent bottles. Suppliers of many laboratory chemicals are now required to label containers with the name of the material, the relevant danger symbol(s) (see fig. 6.5) and a statement of the risks involved in its use. Any chemical covered by the UK Pharmacy and Poisons Act 1933 and later legislation must also carry the warning 'POISON'. The EEC directive warning labels are described in section 6.3 and samples of BDH (British Drug Houses) and Association for Science Education (ASE) labels for a number of common chemicals are shown in fig. 3.9(a). Hazard warning labels are available from most laboratory suppliers or from the Association for Science Education and are recommended for the permanent labelling of all chemical containers (fig. 3.9(b)). The label should carry the IUPAC* recommended name as well as the trivial name if this is still in common use, for example

ETHANOIC ACID
(acetic acid)

or PROPAN-2-OL
(isopropyl alcohol).

* IUPAC is the abbreviation for the International Union of Pure and Applied Chemistry.

The trivial and IUPAC names of a number of common substances are listed in Appendix V.

3.7
Hygiene

The need for continuous attention to hygiene in any type of laboratory cannot be overemphasised. Any spillage of chemicals, battery acid, blood, plasma, serum etc. must be cleaned up at once from the bench, floor or other contaminated equipment and the hands washed thoroughly. Traces of chemicals or other harmful materials can easily be transferred from the fingers to the lips or eyes where considerable damage may result. There is a significant infection risk with virtually any biological material and the appropriate sterilisation procedure should be adopted as required.

Scrupulous cleanliness is essential in any laboratory animal house. Cages should be cleaned and sterilised once a week and floors and benches should be washed regularly with disinfectants. The hands should be washed immediately after handling animals, excreta, cages or water in which fish or reptiles such as terrapins are kept.

3.8
The hazards associated with damaged glassware, cutting tools etc.

Broken glassware, scalpels, razor blades and other cutting tools are the major causes of bloodshed in laboratories (see sections 12.6–12.9). Broken pieces of glass should not be picked up with the fingers, but should be swept up immediately with a dustpan and brush. A piece of plasticine may be used to collect small slivers. All broken glass should be placed in a special receptacle for this purpose and not into the wastepaper basket where it could cut the cleaner's hands. Chipped or cracked glassware should not be used and lengths of glass tubing should always be carried vertically.

Glass tubing may be cut into lengths by making a cut in the tubing at the required place with a glass knife, glass cutter or file. The tube is then wrapped in a cloth and turned over so that the indentation is pointing towards the floor before snapping the glass (fig. 3.10). The sharp edges should be smoothed by heating in a bunsen burner flame or by grinding with corundum. *Safety goggles must be worn when cutting or grinding glass.* Wider tubing may be cut by first making an indentation all the way round the glass with a deeper indentation at the top on which is placed a few drops of cold water. A glass rod is then strongly heated in a flame and placed on the deep cut. A crack usually spreads right round the glass (Fig. 3.11), although frequently more than one attempt may be required. Alternatively a hot wire glass cutting apparatus may be used. Jagged edges may be squared with metal gauze (*Care*: goggles must be worn) and then smoothed in a flame or polished as described above.

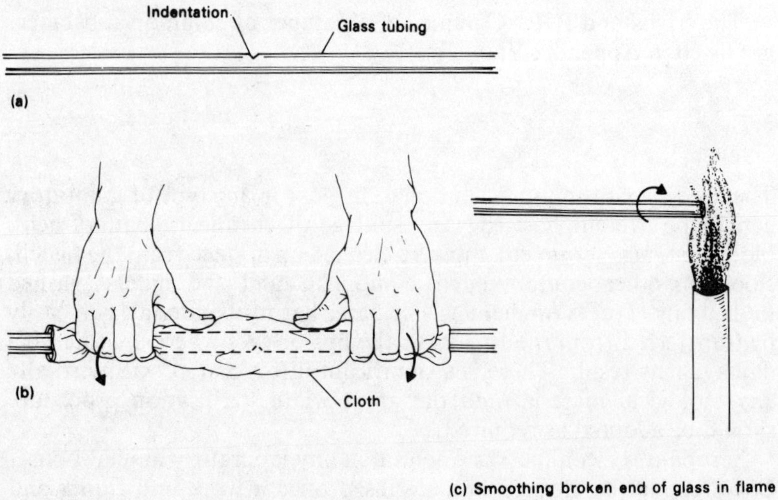

Fig.3.10 Cutting glass tubing

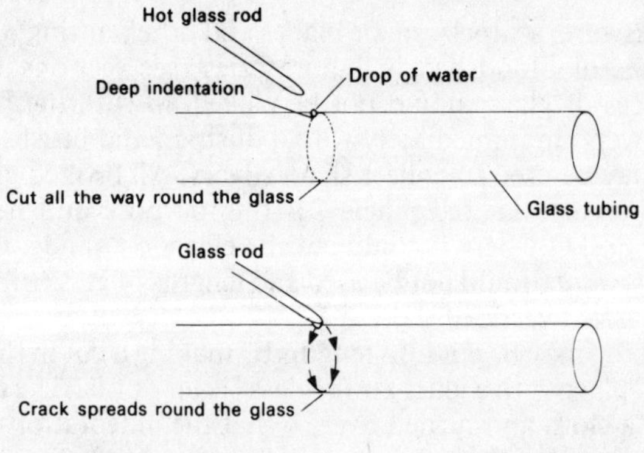

Fig.3.11 Cutting wide glass tubing

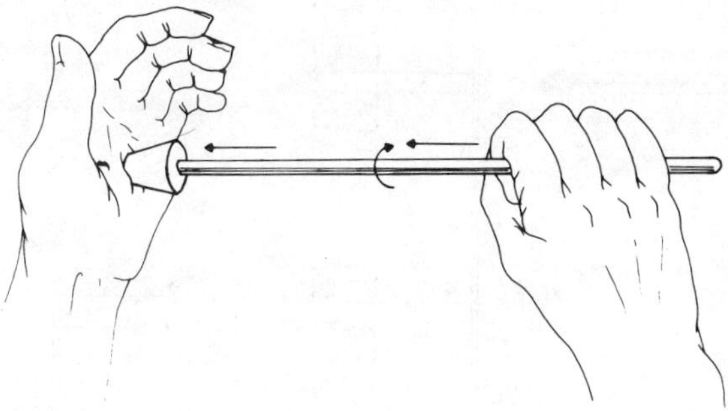

(a) Incorrect method of inserting thermometer or glass tube through a cork or rubber bung

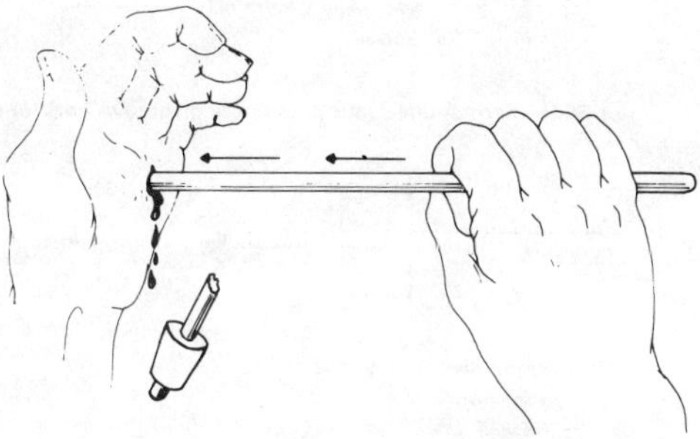

(b) Jagged end of broken thermometer driven into hand

Fig.3.12 The consequences of incorrect insertion of a thermometer or glass tubing into a cork

Accidents often occur when trying to insert a thermometer, or a length of glass tubing or rod through a cork or rubber bung. The glass must not be pushed in from the extreme end as shown in fig. 3.12(a) as this causes the thermometer or tube to break and the jagged end may be driven into the fingers or hand (fig. 3.12(a)). The correct procedure is to lubricate the end of the glass with glycerol, soap or 'Teepol' and then, holding the glass in a cloth next to the cork or bung, gradually ease it into the hole.

Alternatively a cork borer may be used to insert the glass tube or thermometer. The procedure is as follows:

1. Bore the cork in the usual manner with the appropriate size of cork borer so as to leave the hand grip on the side from which the thermometer, for example, is to be inserted (fig. 3.13(a)). *Corks should be wrapped in a piece of paper and rolled under the foot or alternatively in*

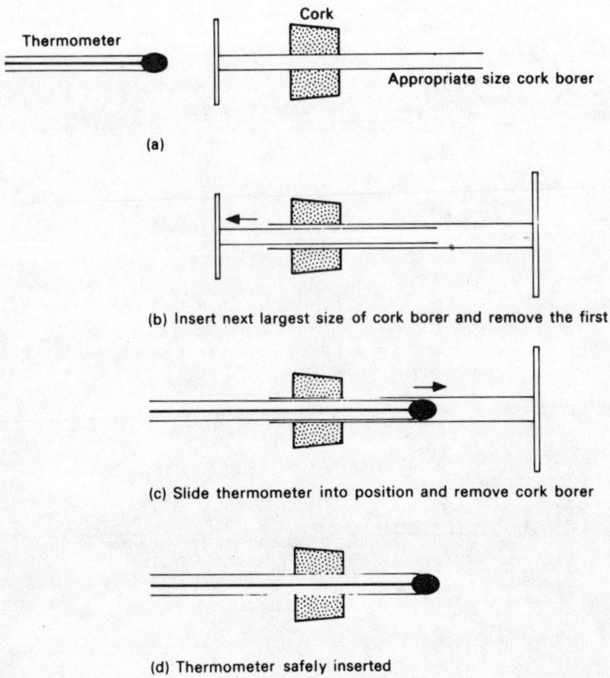

Fig.3.13 Use of a cork borer to insert a thermometer through a cork or rubber bung

Fig.3.14 A cork roller

a cork roller (fig. 3.14) to soften them before they are bored. The hole should be bored half way through the cork from one direction and then, after carefully aligning the lubricated cork borer, a second hole should be bored through from the opposite side to join up with the first (fig. 3.15). The cork borer should then be removed while the sides of the hole are filed smooth.

2. Select the next largest size of cork borer, lubricate it and pass it through the bung on top of the first (fig. 3.13(b)).
3. Remove the smaller cork borer and slide the thermometer into position (fig. 3.13(c)).
4. Withdraw the cork borer (fig. 3.13(d)).

Assignment

Roll and bore a cork as described above and then insert a thermometer or glass rod using either of the recommended procedures.

3.9
Common carcinogens

A number of substances are known to induce cancer in man many months or years after the initial exposure. For example, naphthalen-2-amine (2-naphthylamine or β-naphthylamine) produces malignant tumours of the bladder after a number of years, while bis-chloromethyl

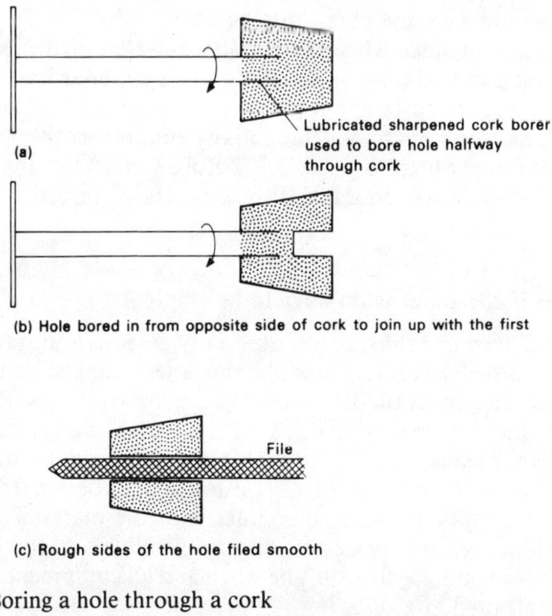

Fig.3.15 Boring a hole through a cork

ether (bis-c.m.e.)—a substance which can be formed spontaneously by the reaction between methanal (formaldehyde) and hydrochloric acid vapour—can cause cancer only a few weeks after exposure. Some substances known as *teratogens* produce mutations in unborn children. Examples of such materials possibly include some herbicides and cortisone and thyroxine. Many substances will induce cancer if implanted in animals over long periods, but may be perfectly safe in normal use; however, any compound for which a cancer risk has been documented should be treated with extreme care and powerful carcinogens should not be used. The Carcinogenic Substances Regulations 1967 prohibit the industrial use and importation of the following compounds or their salts, as well as substances which may contain more than a trace amount of these materials as impurities:

4,4^1-biphenyldiamine (benzidine)
naphthalen-2-amine (2-naphthylamine)
4-aminobiphenyl
4-nitrobiphenyl

The isomer, naphthalen-1-amine, is not known to be carcinogenic, but as it frequently contains the 2-isomer as an impurity, its use is also banned. Other common carcinogenic substances include *o*-tolidine, the chromates, the nitronaphthalenes, the nitrosoamines, chloroethene (vinyl chloride) monomer, 2- and 3-nitrosophenol and some forms of asbestos. It is illegal to prepare, store or use these compounds in school laboratories. Ninhydrin—the reagent used in biological and biochemical tests for the presence of amino acids and proteins—is also carcinogenic and extreme care must be exercised in its use. Aerosol sprays should be avoided wherever possible and if ninhydrin sprays are employed, they should only be used in a fume cupboard.

This list is not exhaustive, but the fact that even quite common substances may have some carcinogenic activity emphasises the necessity of consulting reference literature (see 3.3) before embarking on work with materials when you are ignorant of their potential hazards.

3.10 Precautions if carcinogens do have to be employed

The use of all harmful substances, especially poisons and carcinogens, should be avoided wherever possible and a less dangerous substitute employed if one is available. However, there are occasions when carcinogens have to be used and, as it is rarely possible to fix a safety limit with carcinogens as it is with some toxic substances, extreme care must be exercised in their use. All operations should be carried out in an efficient fume cupboard and all contact with the material should be avoided. Rubber gloves (or better, disposable gloves on top of rubber gloves), lab coat and goggles must be worn and all equipment should be thoroughly cleaned after use.

Assignment

Are carcinogens ever used in the laboratory where you work? What precautions are taken when these substances are employed?

Questions: 3 *Hazards of chemical laboratories*

3.1 Why is it extremely dangerous to direct a jet of compressed air against the skin?
3.2 Why is it wrong to blow dirt or metal filings from crevices or from behind machines with compressed air?
3.3 What precautions should be taken when using Buchner funnels, desiccators and other partially evacuated glassware?
3.4 What are the dangers associated with the use of compressed gas cylinders?
3.5 Oil or grease must never be used on the valve of an oxygen cylinder. Why is this?
3.6 What is a fume cupboard and what is it used for?
3.7 Why is it important to keep poisons locked away and to log them in and out of storage when they are used?
3.8 Why is it important to keep mercury surfaces covered?
3.9 What are the reasons for recommending that the use of benzene as a solvent should be banned?
3.10 Comment on the use of ozonisers to purify air in basements or other enclosed spaces.
3.11 Why is it important to refer to books or other reference material before using appreciable amounts of a substance for the first time?
3.12 Why should a hydrogen sulphide generator (e.g. a Kipp's apparatus) be kept in a fume cupboard?
3.13 Why should bulk supplies of inflammable liquids be stored away from the main buildings?
3.14 Discuss the correct (safe) and incorrect (unsafe) methods of carrying Winchester bottles of concentrated acids or inflammable solvents from the stores.
3.15 Why is it important to always label containers of chemicals? What information should be included on the label?
3.16 What are the reasons for insisting on scrupulous attention to hygiene in chemical, physical and biological laboratories?
3.17 Why shouldn't broken glassware be placed in a wastepaper basket?
3.18 What are the correct and incorrect methods of inserting a thermometer through a cork or rubber bung?
3.19 Why should lengths of glass tubing always be carried vertically?
3.20 What is the procedure for boring a hole through a cork?
3.21 List *four* common carcinogens.
3.22 Why is it necessary to take more stringent precautions with carcinogens than with many toxic chemicals?
3.23 State the precautions which should be employed when carcinogens are used.

Section 4: *The expected learning outcome of this section is that the student should be able to recognise elementary hazards that may be encountered in (a) a physical laboratory and (b) a biological or medical laboratory*

Specific objectives: *The expected learning outcome is that the student:*

4.1 *States that radioactive sources should be kept locked and logged in and out of storage.*
4.2 *Recognises a u.v. source.*
4.3 *States that, as a minimum precaution, u.v. sources should be viewed through glass.*
4.4 *Recognises the potential hazard of laser light.*
4.5 *States that the principal hazards of working in a biological laboratory are infection and disease and describes how these dangers may be minimised.*
4.6 *Lists the methods employed for sterilising apparatus used for microbiological experiments.*

4.1
Radioactive sources

A radioactive isotope is a substance which undergoes spontaneous disintegration to form atoms of a different element. This disintegration is accompanied by the emission of radiation of which there are three types, designated by the first letters of the Greek alphabet (see table 4.1). All three types of radiation can be detected by a Geiger counter or by their effect on photographic film. The rays differ in the extent to which they will penetrate materials or produce ionisation. The energy of the radiations absorbed when these ionising rays pass through living organisms can cause immense damage to the body tissues. Actively dividing cells, such as those which produce the red blood corpuscles, are particularly susceptible. The effects of radiation on the reproductive cells are cumulative and genetic damage may result. Great care must therefore be taken in the use, storage and disposal of such materials.

Table 4.1 *Types of radiation*

Radiation	Nature	Relative ionising power	Penetration
Alpha(α)-rays	A stream of α-particles (mass	Most ionising	Very short range—a few cm of air or thin

Table 4.1 *(Continued)*

Radiation	Nature	Relative ionising power	Penetration
	= 4 a.m.u.; charge +2)		sheets of card or metal foil.
Beta(β)-rays	A stream of electrons (rest mass = 1/1840 a.m.u.; charge −1)	$< \alpha$	Stopped by a few mm of lead. High energy β-rays will penetrate thin aluminium sheet.
Gamma(γ)-rays	Uncharged radiation similar to X-rays but of shorter wavelength (see fig. 4.2)	Least ionising	Most penetrating—about 100 times more penetrating than α-rays. Will pass through steel, lead, flesh etc. Every 1 cm of lead approximately halves the intensity of γ-radiation.

(a) *Units of radiation*

The *curie* (symbol Ci) is the commonly accepted unit for representing the activity of a quantity of radioactive material. It is equal to a rate of 3.700×10^{10} disintegrations per second which is approximately the activity of 1 g of radium. The curie is an extremely large unit, so the millicurie (mCi) and microcurie (μCi) (see Appendix I), are commonly employed for most practical purposes. The specific activity of a material is usually expressed in curies per gramme (Ci g^{-1}).

The curie is slowly being replaced by the SI unit, which is called the *becquerel* (symbol Bq), where

$$1 \text{ becquerel} = 1 \text{ disintegration per second} = \text{approx. } 27.03 \times 10^{-12} \text{ Ci}$$

so that

$$1 \text{ Ci} = 3.7 \times 10^{10} \text{ Bq}.$$

A small demonstration source for college use is about 1 μCi or 3.7×10^4 Bq.

The unit of absorbed energy is the *gray* (Gy) which is equal to an absorbed energy of one joule per kilogram. This unit supercedes the *rad*, where

$$1 \text{ Gy} = 100 \text{ rad}.$$

The damage produced in biological systems by the absorption of radiation depends on a number of factors such as the dose rate during irradiation, the duration of the exposure and the type of radiation. Heavy particles such as protons, neutrons and alpha-particles produce more damage than X-rays, gamma-rays or beta-rays. The SI unit is the *sievert* (Sv) where

$$1 \text{ Sv} = \text{absorbed energy (Gy or J kg}^{-1}) \times \text{a factor which depends on the type of radiation absorbed.}$$

However, the rem or millirem (mrem) is still in common use and maximum permissible doses are usually specified in this unit. A number of examples are given in table 4.2. Pupils in schools and colleges should not receive more than 50 mrem in any one year. The maximum permitted dose for specialised workers in industry (5 rem p.a.) is considerably higher than this, but such workers are subjected to regular monitoring and health checks. The radiation received is usually measured by a special photographic film mounted in a badge worn by the worker. The holder contains several metal foils so the blackening of different areas of the film identifies the type of radiation and provides information about its energy and the dose absorbed. These films are replaced and processed every two or three weeks.

Table 4.2 Radiation doses

Activity	Approximate dose
Chest X-ray	100–200 mrem
Watching colour TV at a range of 2 m	approx. 0.5 mrem h^{-1}
Natural cosmic radiation at sea level	30 mrem p.a.
Natural radioactivity in rocks	50 mrem p.a.*
Small amounts of radioisotopes, e.g. potassium-40 in the body	25 mrem p.a.*

* The total radiation absorbed from natural sources amounts to approximately 8 rem in a lifetime.

A dose of more than 400 rem over the whole body would probably be lethal, while 200 rem over a short period of time would be likely to cause leukemia. Radiation affects the bone marrow, blood, spleen and lymph nodes. It can cause cancer and—owing to the decrease in the number of white blood cells—it drastically reduces the body's ability to resist infection.

(b) *Legal restrictions in the use of radioactive sources*
The use of radioactive materials in schools and colleges is strictly controlled and approval of the Secretary of State is required for equipment with an activity greater than 3.7 kBq (1 μCi). The legal requirements and a code of practice are summarised in the Department of Education and Science Administrative Memorandum 2/76, *The Use of Ionising Radiations in Educational Establishments.*

Industrial sources are frequently more than a million times stronger than the demonstration sources in schools and colleges. Extensive lead or steel shielding is required and specialised equipment and techniques are needed to work with such materials.

(c) *Recommendations when using low power sources*
The following guidelines are suggested for experiments using low activity radioactive materials:

1. Radioactive substances must not be allowed to come into contact with the skin. Always wear protective gloves (disposable polythene gloves over a rubber pair are convenient) and always wear a lab coat or other suitable protective clothing. Protective clothing should be monitored after use and left in the laboratory. Powdered chemicals should be transferred in a fume cupboard or isolation or manipulator glove box (see fig. 4.1) with a spatula and small sources or chromatography paper bearing radioactive substances should be handled with forceps. Safety bulbs (see fig. 9.21) must be used when pipetting radioactive substances. Do not lick any labels.
2. Do not eat, drink or smoke in the laboratory.

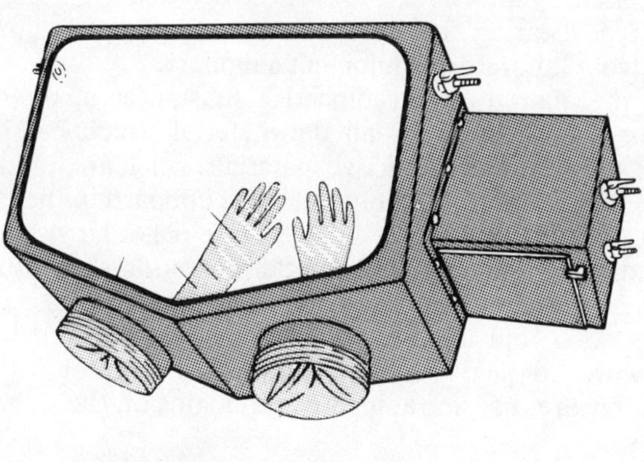

Fig.4.1 An isolation or manipulator glove box

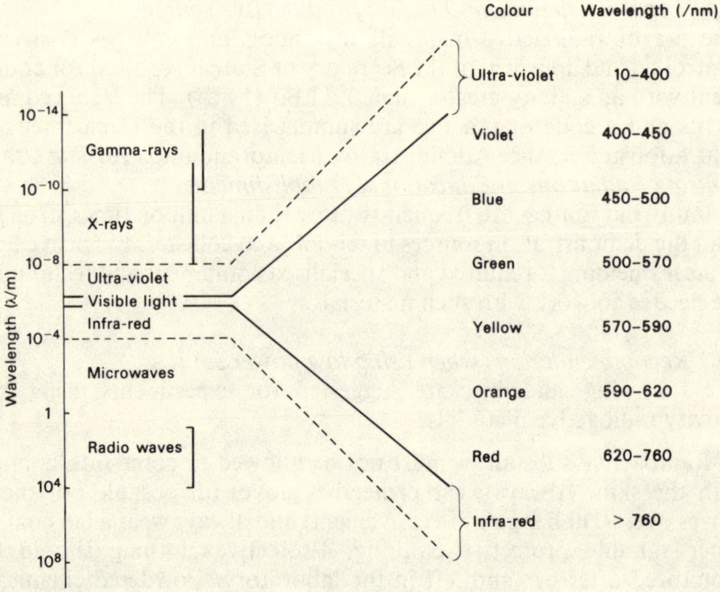

Fig.4.2 Ultra-violet light and the electromagnetic spectrum

3. Confine any work with radioactive materials to one area of the laboratory such as a fume cupboard and always set up warning signs.
4. Glassware for use with radioactive materials should be kept solely for this purpose and stored separately to guard against accidental contamination.
5. All bottles containing radioactive material should be labelled with the recognised symbol (see sub-section 6.1 and fig. 6.1). Rolls of adhesive type bearing this symbol and the word 'Radioactive' are available from laboratory equipment suppliers.
6. Check the quantities of radioactive substances at the end of the experiment and account for all the material which has been used. Return all sources and radioactive materials which are not disposed of to a securely locked and suitably shielded cupboard in the stores. This cupboard should be properly labelled (see 6.1) and used solely for storing radioactive materials. A special notebook should be kept for logging these radioactive sources in and out of storage.
7. Always wash your hands thoroughly using disposable towels after working with radioactive materials and then monitor with a Geiger counter to ensure that no radioactivity remains on the skin.

Assignment

Compile a list of the radioactive sources and their activity which are located in the laboratories at work and in the college. Where are these materials stored and what precautions are taken to shield people from the radiation produced? Discuss the additional risks presented by the storage of these materials in the event of a fire.

4.2 Ultra-violet sources

Ultra-violet light is the name given to the portion of the electromagnetic spectrum immediately beyond the visible region (fig. 4.2). It has a shorter wavelength than visible light. Hydrogen lamps, deuterium lamps and other sources of ultra-violet light are used for a variety of purposes in physical, chemical and biological laboratories. These applications include spectroscopy, the sterilisation of instruments, the initiation of chemical reactions and the detection of purines and other substances which fluoresce when irradiated with u.v. light in chromatography and other separation techniques. Ultra-violet sources may be recognised by their intense pale blue illumination. Large amounts of u.v. light are produced during arc welding and proper eye protection (see sub-section 4.3) must always be worn when watching or carrying out such operations.

4.3 Protection against ultra-violet radiation

Exposure to ultra-violet light can cause serious damage to the skin and to the outer layers of the eye. The symptoms after slight exposure are the same as those of severe sunburn, but with increasing exposure, intense pain, irritation and watering of the eyes, a feeling of having sand in the eyes and temporary loss of vision result. Conjunctivitis frequently occurs some time after exposure and may last for several days. An excruciatingly painful condition known as 'eye flash' may also be produced by looking at an exposed source of u.v. light.

Ultra-violet lamps must always be properly shielded and approved goggles should be worn by anyone working with or close to the source. As a minimum precaution the source should be viewed through 'Pyrex' glass as this material absorbs ultra-violet light and is thus largely opaque to radiation of this wavelength. The skin of the hands, face or forearms can be protected by thick cloth or by suitable barrier cream if necessary. The room should be well ventilated to guard against the dangers of inhaling excessive amounts of ozone, which can be produced by the effect of ultra-violet radiation on the oxygen in the air.

Assignment

Are u.v. sources used in the laboratories at work or in college? What are they used for and what protective measures are taken to guard the operator against the effect of u.v. light?

4.4
The potential hazard of laser light

Laser light is an extremely intense form of electromagnetic radiation. Its name is derived from the initial letters of Light Amplification by Stimulated Emission of Radiation. Lasers are now used to a rapidly increasing extent in industry, research laboratories and in schools. Examples of their industrial use include drilling holes in metals and for cutting cloth and metal sheet. Their dangers stem from the enormous concentration of the energy of the laser into a very small area. These extremely short, concentrated pulses of light can cause severe eye damage. Both the cornea and the retina may be affected and permanent blind spots can be produced at the points where the lens of the eye brings the rays to a focus, thereby destroying the tissues at the back of the eye. This focussing effect of the lens of the eye increases the intensity of the laser light by a factor of about a million (10^6), so laser light with an energy density as low as 10^{-8} J cm^{-2} at the pupil of the eye can be dangerous. In practice, the maximum safety value is set at one-tenth of this figure. Additional dangers in the use of the laser stem from the very high voltages required to operate its source.

Only lasers of low power are permitted in schools. The code of practice governing their use is described in the Administrative Memorandum AM 7/70 of the Department of Education and Science, *The Use of Lasers in Schools and Other Educational Establishments.* Teachers and technicians are recommended to refer to this publication before carrying out any work with lasers. Considerably more powerful lasers are used in industry and research. Again, strict adherence to the code of practice (see 9.17 and BS 4803 : 1972, for example) is essential to avoid accidents. The following general rules must always be observed:

1. Never look along a laser beam or expose any part of the skin to it. Remember damage is caused by both reflected and direct laser radiation and any surface which reflects light will also reflect a laser. Block out any reflected laser light with opaque matt grey screens. Always wear the correct type of goggles for the particular laser being used. A helium–neon laser, for example, gives a red beam. Some lasers generate radiation in the ultra-violet and infra-red regions of the spectrum as well as producing visible light and the appropriate precautions for protection should be taken.
2. Never operate lasers in the dark. The room should always be brightly lit to avoid enlargement of the pupil.
3. Warning notices (see fig. 6.2) should be set up and the area should be fenced off when lasers are in operation.
4. The laser source should be fixed rigidly to the bench so that the direction of the beam cannot be altered by accidental movement.
5. Never align a laser while it is switched on. This alignment should be carried out optically before turning on the source.

6. Report any incident of accidental exposure to radiation. In serious cases or if in any doubt, seek medical advice.
7. Keep the laser in a securely locked storeroom when it is not in use.

4.5
Safety in biological laboratories

Laboratories in hospitals, veterinary establishments, pharmaceutical companies, departments of forensic science and the biology department of schools, colleges and universities are all concerned with the examination of living organisms and of animal, human and plant tissues or of specimens taken from these sources. The particular hazards of the biological laboratory are infection and disease and frequently the common hazards of all chemical laboratories are also present. Viral hepatitis is a major hazard of work in hospital laboratories where the disease is frequently contracted from blood, urine and other samples sent in for tests and laboratory examination. Infected matter can be inhaled or ingested in the form of a fine spray or aerosol produced during stirring, centrifugation etc. or as a dust from dried material. It may also find its way into the body through cuts, scratches and other breaks in the surface of the skin. Specialised techniques are required for the manipulation of high risk materials. These include the use of isolation boxes for transfering radioactive

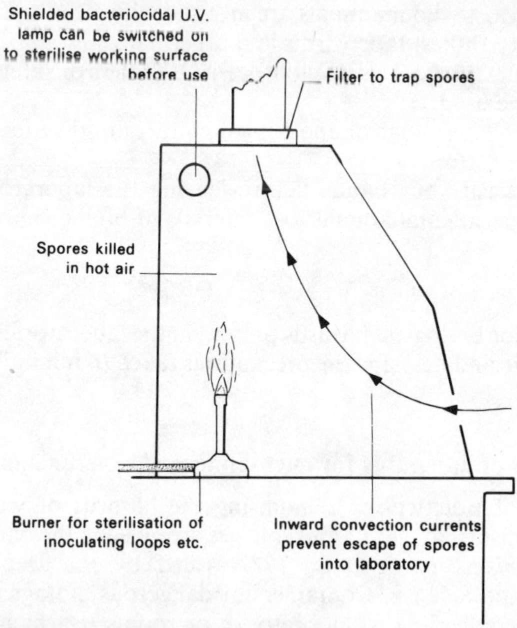

Fig.4.3 A transfer chamber (see also Fig.7.1).

substances (see fig. 4.1) and of a transfer chamber (see figs. 4.3 and 7.1) for plating out and sub-culturing micro-organisms. Other dangers in biological laboratories come from the keeping of experimental animals and the possibility of stings, bites and scratches. Great care must also be taken in microbiological experiments, particularly with pathogenic (disease producing) organisms. Many micro-organisms which are normally harmless, such as *E. coli* in the intestines, can produce disease in a different habitat. *E. coli*, for example, can cause septicaemia if it finds its way into the bloodstream.

The dangers of working in such laboratories will be minimised if the following precautions are taken:

1. Wash all dissection instruments in disinfectant after use.
2. Never eat in the laboratory and do not eat seeds or parts of plants provided for study as they may have been treated with toxic fungicides. Never store milk and other foodstuffs in the same refrigerator as dissected specimens, serum, microbiological cultures or other biological material.
3. Never take wild birds or mammals into the laboratory as they can carry and transmit diseases fatal to man. Examples of such diseases include psittacosis from wild birds, recaptured canaries or budgerigars and Weal's disease from wild rodents. Monkeys and other primates may be carriers of hookworm, rabies and B-virus infection, while terrapins and laboratory rats are a frequent source of salmonella poisoning. Most wild mammals are infected with fleas and other pests and can carry flukes, tapeworms and other parasites.
4. Never pipette solutions by mouth. Pipette fillers or safety bulbs (see 9.21) must be used.
5. Soak all bacteriological and fungal cultures in disinfectant before disposal (see 8.1(c)).
6. Always wash your hands before leaving the laboratory or after handling experimental animals or materials of biological origin.

Assignment

List the major biological hazards present in the laboratories at work or in the college and describe the precautions taken to minimise the risks.

4.6
Sterilisation of apparatus for microbiological experiments

The codes of practice for minimising the hazards of working with micro-organisms in schools and colleges are given in the booklet *Use of Micro-organisms in Schools, 1977* issued by the Department of Education and Science. Containers of dangerous biological materials and the doors leading to laboratories or rooms in which work with pathogenic micro-organisms is carried out should be labelled with the *biohazard warning symbol* (see fig. 6.4). The following methods are employed to sterilise apparatus for microbiological experiments.

Glassware (e.g. Petri dishes, pipettes, flasks, test-tubes and syringes) Stopper the flasks and tubes loosely with cotton wool and wrap pipettes and syringes in brown paper or metal foil before sterilising in a hot air oven at 160 °C for 1 hour.
Innoculating loops, mouths of culture tubes, slides, cover-slips and forceps points Pass through a flame before use.
Contaminated floors and benches Wash with 3% 'Lysol' solution (**Care**: Do not allow this solution to come into contact with your skin) or other suitable disinfectant solution.
Culture media, rubber washers and bottle caps Heat for 30 min under pressure in an autoclave or pressure cooker.
Disposal of old cultures See 8.1(c).

Questions: 4 *Hazard of physical and biological or medical laboratories*

4.1 Name *three* types of ionising radiation. Compare and contrast their properties.
4.2 Describe *two* methods of detecting ionising radiation.
4.3 Why is it considered to be unsafe to lick labels for bottles containing radioactive materials?
4.4 Why is it important to lock radioactive materials away and to log them in and out of storage?
4.5 What is ultra-violet light? What are u.v. sources used for?
4.6 What are the potential hazards of (a) ultra-violet light and (b) lasers? How are these dangers minimised?
4.7 Why is it advisable to fix a laser firmly to the bench before carrying out an experiment?
4.8 Describe three precautions which should be taken to minimise the risks of infection when working in a medical laboratory.
4.9 List the methods used to sterilise apparatus used for microbiological experiments.
4.10 What is a pathogen?
4.11 Why is it important to cover cuts on the skin when working in a biological laboratory?
4.12 Why is it unwise to take rodents, birds and other wild animals into a school laboratory?

Section 5: *The expected learning outcome of this section is that the student should know the basic precautions to prevent fire and the action to be taken in the event of fire*

Specific objectives: *The expected learning outcome is that the student*:

5.1 Lists common fire hazards in the college laboratories and at work.
5.2 States the purpose of a fire door.
5.3 Lists the main types of fire extinguisher.
5.4 States the range of application of each of the main types of extinguisher.
5.5 Knows how to use each of the types listed under 5.3
5.6 Recognises the need for knowing the position of fire alarm buttons both at college and work.
5.7 States for both college and work situation escape routes from the building and assembly points outside.
5.8 Recognises the need, if time permits, to close doors and windows on evacuating the building.

5.1
Common fire hazards

(a) *The fire triangle* In general three factors must be present or come together for a fire to occur and continue burning:

1. a fuel, i.e. something to burn;
2. a source of oxygen to support combustion;
3. heat to ignite the fuel and sustain burning.

As a simplification in fire prevention and control these factors are represented diagrammatically as the three sides of what is known as the 'fire triangle' (fig. 5.1). The removal of any one of the three sides causes the triangle to collapse and indicates that fire outbreak is impossible or a fire which has already started cannot continue burning.

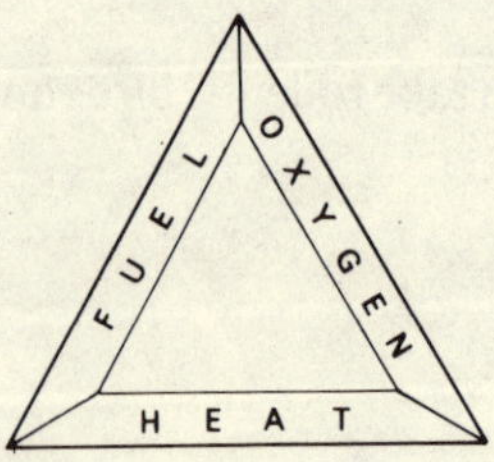

Fig.5.1 The fire triangle

This simple principle should be borne in mind when attempting to extinguish a fire or when planning the construction, organisation and layout of a laboratory, workshop or store so as to minimise the possibility of fire outbreak and to reduce its possible effect should a fire occur. Most fire extinguishers act by excluding the oxygen supply from the fire and by reducing the fire temperature to a level below which burning will not continue. Some types of extinguisher for special fires act by interfering with the chemical processes which occur during the combustion.

The oxygen supply for a fire generally comes from the air in which it is present in about 20% by volume. Other potentially dangerous sources are chlorates, peroxides, permanganates, nitrates, dichromates and other powerful oxidising agents. These substances must be stored away from organic solvents and other highly flammable materials.

Fire is a self sustaining combustion process which takes place in the vapour phase, producing heat, smoke and incandescence. Wood, paper, plastic, coal and other solids usually have to be heated to a sufficiently high temperature to give off an inflammable vapour before they will ignite, but many organic solvents produce sufficient vapour even at very low temperatures to present a serious fire hazard. Gases, finely divided powders or dusts, and the vapour from flammable solvents may be ignited by the short burst of energy from a spark.

The cost of the damage caused by fire in the United Kingdom comes to more than a million pounds each *day*, and this does not take into account the deaths, injury and suffering which the fire may cause. The ten most likely ignition sources for large fires are: smoking materials (cigarettes, matches, pipes etc.); misused or faulty electrical installations; mechanical heat or sparks; naked lights; oxyacetylene welding and cutting equipment; malicious or intentional ignition; children playing with fire; gas appliances and installations; oil appliances and installations; deliberately burning rubbish. The fuel these sources of ignition are most likely to set alight are: waste and rubbish; paper, cardboard, plastic foam and other packing and wrapping materials; textiles; flammable liquids; electrical insulation and combustible elements in the structure and fittings of the building. The essence of fire prevention is to keep the fuels and sources of ignition apart. Most laboratory fires could be avoided by following this simple rule.

Assignment

Compile a list of the major combustible materials in the college laboratory and at work which would act as a fuel in the event of the fire. To what extent would the replacement of wood, plastic and other inflammable materials in the structure and fittings of the laboratory minimise the risk and extent of a fire?

(b) *Classification of fires*
Fires are classified into four types according to the nature of the material undergoing combustion (see table 5.1). This classification is important in firefighting as the choice of an extinguisher best suited to deal with a particular outbreak is determined by the nature of the material which is burning. The classification code (European Standard EN 2 : 1972) outlined in table 5.1 has been adopted in Britain.*

Table 5.1 *Classification of fires*

Class of fire	Type or nature of fire
Class A	Fires involving carbonaceous materials, e.g. wood, cloth, paper, rubber. These materials incandesce and can produce glowing embers. Flammable gases and vapours are obtained by destructive distillation on heating.
Class B	Fires involving flammable liquids, e.g. petrol, oil, alcohol and many other organic solvents.
Class C	Fires involving flammable gases, e.g. methane, propane, hydrogen, ethyne (acetylene) and butane.
Class D	Fires involving flammable metals, e.g. sodium, potassium, calcium, magnesium and other combustible metals or their hydrides.

Virtually all laboratories have a mains supply to power electrical appliances and equipment and many, if not all, the types of combustible materials listed in table 5.1 may be present. Fires of all four classes are thus a possibility with the added complication and hazard of electricity. All technicians should be aware of the potential hazards and should never underestimate the possible effects of a fire outbreak.

Assignment

List the fire hazards in (1) the college laboratory and (2) at work. In each case classify the type of fire which would result.

* This replaces the earlier standard which defined a Class C fire as one involving electrical equipment.

Note: Although the prefix 'in-' often imparts the negative or opposite meaning to an adjective, as in the words inaccurate, inadequate or independent for example, this does not apply to the words 'flammable' and 'inflammable'. Both these adjectives have essentially the same meaning, i.e. easily kindled or that which may be set on fire. It has been suggested that only the word 'flammable' should be used so as to minimise the possibility of mistakes. However this would probably increase the danger of other mistakes occurring with liquids labelled as 'inflammable' as the word 'flammable' became the generally adopted form. It is essential therefore for any technician or anyone responsible for the training of those working in laboratories to know that the words are mutually synonymous and this is the procedure which has been adopted in this book.

(c) *Flammable (or inflammable) liquids*
There are wide differences in the ease with which combustible materials will ignite and burn. For example, wood or liquid paraffin can be heated to quite a high temperature before it bursts into flame, while petrol or carbon disulphide may be ignited if its vapour comes into contact with the hot glass surface of an electric light bulb. Three figures are important when assessing the fire hazards of an inflammable liquid:

1. its flash point;
2. its ignition temperature;
3. its explosive limits.

The values of these quantities for a number of common liquids are listed in table 5.2.

Table 5.2 *Flash points, ignition temperatures and explosive limits of common substances*

Liquid	Flash point (/°C)	Ignition temperature at atmospheric pressure in air /°C	Explosive limits in air at 25° (%)	
Benzene	− 11	562	1.8	8
Carbon disulphide	− 30	100	1	44
Diethyl ether	− 45	180	1.85	48
Dioxan	12	180	2	22
Ethanal (acetaldehyde)	− 38	185	4	57
Ethanoic (acetic) acid	43	426	4	16
Ethanol	12	423	3.3	19
Ethyl ethanoate (acetate)	− 4.4	427	2.5	11.5

Table 5.2 (Continued)

Liquid	Flash point (/°C)	Ignition temperature at atmospheric pressure in air /°C	Explosive limits in air at 25° (%)	
Hexane	– 23	260	1.2	7.5
Methanol	10	464	7.3	36.5
Petroleum ether (lower fractions)	– 17	from 290	1	6 (approx.)
Propanone (acetone)	– 18	538	2.5	13

1. *The flash point of a liquid* is the temperature above which a liquid gives off sufficient vapour to form an inflammable mixture with air. It is thus the temperature above which the liquid will ignite in the presence of a flame.
2. *Ignition temperature* This is the minimum temperature required to initiate (or maintain) self sustained combustion of a fuel in air at atmospheric pressure. Ignition temperatures are strongly dependent on the oxygen content of the atmosphere in contact with the combustible material. For example, cloth which will only smoulder slowly in air containing 20% of oxygen by volume burns fiercely in an atmosphere in which the oxygen content has been increased to 25%. An increase in pressure or even a slight increase in the proportion of oxygen in the air can have a marked effect on the ease with which a substance will ignite, thereby resulting in an enormously increased fire risk. Oil, grease and other organic substances must never be used on the valves of oxygen cylinders as the reactions of the gas with the materials is explosively violent (see section 3).
3. *Explosive limits* This is the name given to the range of percentage by volume of the vapour of the flammable component in a mixture with air within which there is a risk of explosion. Some substances, such as carbon disulphide, ether and ethanal (acetaldehyde), have such low ignition temperatures and flash points that an explosive mixture can be ignited by the hot glass of an electric light bulb or by sparks from nylon clothing charged with static electricity. Inflammable liquids can also be ignited by sparks from electrical switches (such as that inside a laboratory refrigerator or for electric lights) and by the high temperatures generated by friction in drilling or by badly lubricated bearings.

Assignment

Look up the flash points, ignition temperatures and explosive limits of the common flammable organic solvents you use in the laboratory at work. Do these figures merit taking greater care when using these liquids in the future? What precautions do you suggest?

5.2 Fire doors

A fire, once started, can spread very quickly by the movement of flames to adjacent combustible materials. Radiation of heat across air spaces can rapidly heat other combustible material to its ignition temperature when it too bursts into flame. The upward movement and growth of a fire occurs by a similar process and (particularly) by convection. A column of hot gases and smoke from the fire rises until it meets the ceiling or some other obstruction and then mushrooms out until it reaches a lift shaft, duct, staircase or other open aperture which will enable it to continue to rise. The continued flow of hot gases against a ceiling made of combustible material heats it above its ignition temperature and it too will burst into flames. This in turn sets light to combustible roof linings and to the wooden floors and joists of the floor above and by a combination of convection, conduction and radiation continues the spread of the fire. The effect of high temperatures on concrete, plaster, metal, glass and other noncombustible parts of the building can weaken their structure and allow the fire to spread and in some cases causes the building to collapse.

The most effective method of limiting the spread of fire is to subdivide the interior of a large building into a number of small separate compartments by means of fire resisting walls, floors and doors. Fire doors (or smoke doors) limit the spread of fire by presenting a physical barrier to the flames and, in the case of an outbreak in a small room, by limiting the supply of oxygen to the fire. They also contain the smoke and fumes from the fire and give people the opportunity to escape from the building before it is too late.

Smoke and fumes are the principal causes of the thousand or so deaths in fire which occur every year in Britain. Many of these fumes are extremely toxic, for example the combustion products of the polyurethane foams commonly used in furniture contain hydrogen cyanide. Carbon monoxide is also present in the gases from virtually any fire and—as this gas is odourless and has virtually the same density as air—it quickly spreads and may remain undetected until it is too late. A small amount of carbon monoxide impairs a victim's judgement and continued inhalation quickly produces unconsciousness and death. Many of the people killed in fires die of poisoning in the first minutes of an outbreak and occasionally before they are even aware that a fire has broken out.

Fire doors are fitted with self closing hinges and open towards the exits from the building. They must be kept closed at all times and the practice of wedging a fire door open in hot weather should be firmly discouraged. It is important to remember that any fire will deplete the supply of oxygen in a room as well as producing quantities of poisonous fumes and gases. Care should be taken to ensure that all rooms are well ventilated before returning to a building after a fire.

Assignment

State the functions of a fire door. Draw a plan to show how two floors of the college or the interior of the building you work in are separated into compartments by the presence of fire doors.

5.3
The main types of fire extinguisher

The main types of fire extinguisher are as follows.

(a) *Water*
This may be provided by a hose or by a soda-acid or carbon dioxide expelled water extinguisher. *The soda-acid extinguisher* contains about 9 litres of aqueous sodium hydrogen carbonate (sodium bicarbonate) solution. The extinguisher is operated by breaking the container of sulphuric acid in the top (fig. 5.2) and the carbon dioxide liberated by reaction of the acid with the sodium hydrogen carbonate forces a jet of solution through the nozzle. In the *carbon dioxide expelled water extinguisher* (see fig. 5.3) which has now largely replaced the soda-acid extinguisher, the water is forced through the jet when the operator punctures a small cylinder of carbon dioxide. These two fire extinguishers cannot be turned off until they are exhausted; however some firms are now producing models with a squeeze grip control to regulate the discharge of the extinguishing medium. Water extinguishers are painted red for easy identification. The fire is extinguished by the cooling action of the water (see 5.1(a)).

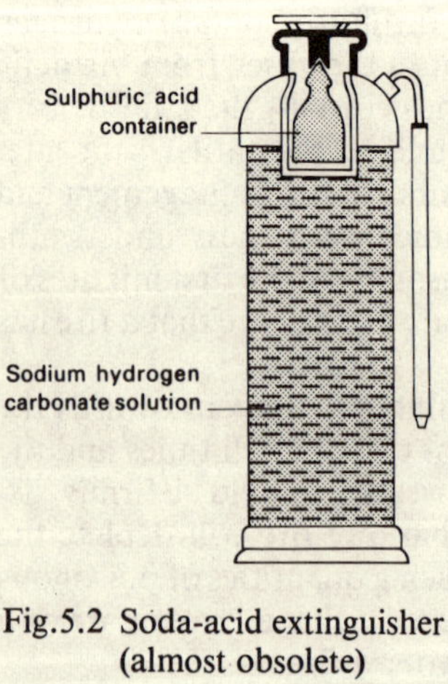

Fig.5.2 Soda-acid extinguisher (almost obsolete)

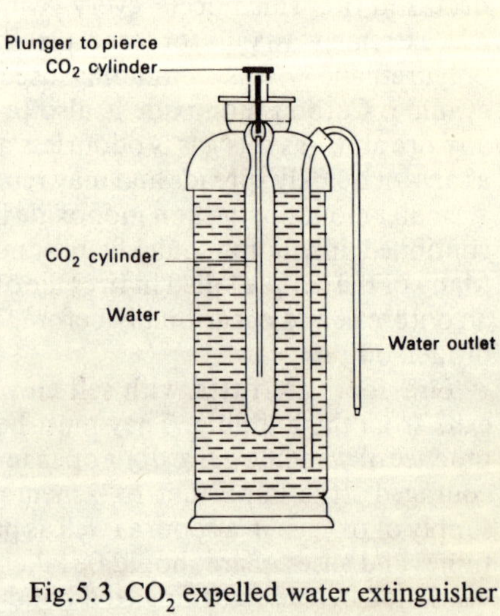

Fig.5.3 CO_2 expelled water extinguisher

(b) *Carbon dioxide extinguishers*

Carbon dioxide (CO_2) extinguishers consist of a cylinder of carbon dioxide at high pressure (fig. 5.4). The gas is expelled through a plastic horn when the valve or trigger is released. The extinguisher can be turned off when required. The jet of gas expelled is very cold and freezes the water vapour in the air and produces a cloud of solid carbon dioxide. The plastic horn should not be touched during or after operating the extinguisher because of this intense cold. Carbon dioxide is denser than air and the extinguisher acts partly by its cooling effect but mainly by excluding oxygen from the fire. A black colour coding is used for carbon dioxide extinguishers.

Fig.5.4 Carbon dioxide extinguisher

(c) *Foam extinguishers*

These work in a similar manner to the soda acid extinguisher but contain an aqueous solution of a gelatinous material and foaming agent so that the bubbles of gas become trapped and form large amounts of a stable sticky foam. The extinguisher acts by excluding oxygen from the fuel and to a small extent by its cooling effect. To be successful an unbroken layer of foam should be spread over the fire. Foam extinguishers are painted white for easy recognition. Like the product of the soda–acid extinguisher, foam can be very messy to clean up after a fire.

(d) *Vaporising liquid extinguishers*

These act by blanketing the fire in a layer of dense vapour from a non-inflammable liquid to exclude oxygen from the combustion site. *Tetrachloromethane* (*carbon tetrachloride*) was once used in these extinguishers but as it yields the toxic gas carbonyl chloride among its combustion products it has now been replaced by bromo-chlorodifluoromethane, $CBrClF_2$, b.c.f. (Carbonyl chloride is another name for phosgene—the poisonous gas used in World War I.) BCF is supplied in pressurised containers (colour code: green). The gas is

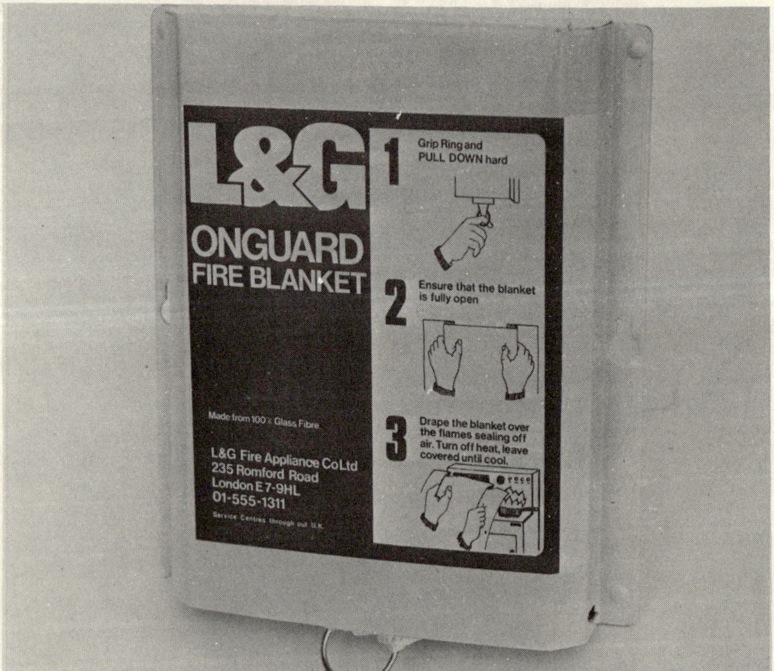

Fig.5.5 A glass fibre fire blanket

almost inert but yields toxic and irritant combustion products. This type of extinguisher is widely used in industry, but is not recommended for college use.

(e) *Powder extinguishers*
These act by blanketing the fire in an inert powder and thus extinguish the fire by excluding oxygen. The principal examples are *dry sand* and *dry powder extinguishers*. The latter propels a layer of sodium hydrogen carbonate over the fire when the pressure of the carbon dioxide in the cylinder is released. The heat of the fire also liberates carbon dioxide from the powder by the endothermic reaction:

$$2NaHCO_3 \xrightarrow{\text{heat}} Na_2CO_3 + H_2O + CO_2.$$

Fig.5.6 Principal types of fire extinguisher: water extinguisher (left); BCF extinguisher (right).

A dry powder extinguisher (colour code: blue) cannot be switched off once it is turned on until the cylinder is exhausted.

(f) *Blankets*
Fire blankets act by excluding oxygen from the air by the smothering action of the non-combustible material of which they are made. Light, flexible woven glass-fibre blankets (fig. 5.5) are now used in preference to the heavy, cumbersome asbestos blankets. Asbestos blankets are no longer used because of the danger of asbestosis and cancer from inhaled fibres of the material.

The principal types of fire extinguisher are shown in figs. 5.6 and 5.7.

Assignment

Note the type, number and location of the fire extinguishers in the college laboratory and at work. List the colour codings used to identify the different types of extinguisher.

Fig.5.7 (a) A foam fire extinguisher

Fig.5.7 (b) A powder fire extinguisher

5.4
Range of application of the main types of fire extinguisher

The types of fire (see 5.1(b)) which may be extinguished by the various fire extinguishers are described in table 5.3 and summarised in table 5.4.

A minimum requirement for most laboratories is a carbon dioxide extinguisher, a fire blanket, a water hose and a bucket of sand. Additional fire fighting equipment, such as BCF and foam extinguishers and automatic drenching systems, is frequently necessary on chemical plant or in specialised industrial laboratories. This should be installed after consultation between the fire brigade and the firm's safety officers.

Sand buckets should always be kept filled with dry sand and not used as a receptacle for waste paper and cigarette ends. A long handled scoop should be provided. Dry sand may be used to smother sodium, potassium and phosphorus fires to allow time to evacuate the building. The fire brigade must be called for fires of this type (except the most trivial) as none of the commonly available extinguishers is effective against burning sodium or potassium. Carbon dioxide, for example, is split by the burning metal into oxygen and soot.

Table 5.3 *Main types of fire extinguisher and their applications*

Type of extinguisher	Application	Disadvantages/ limitations	Notes
(a) Water	Extinguishing burning wood, paper, timber etc. (Class A fires).	Must not be used on: (i) electrical fires or fires close to electrical supply; (ii) solvent fires as it causes them to spread; (iii) sodium, potassium, magnesium and calcium fires.	Not suitable for general laboratory use.
(b) Carbon dioxide	Suitable for virtually any small fire, especially Class B fires (burning liquids) and for those involving electrical equipment.	Has virtually no cooling effect and the fire may reignite as the gas disperses. Ineffective on sodium fires. Should be used with care in confined places as the carbon dioxide concentration required to suffocate people is lower than that required to extinguish a fire.	Best extinguisher for general laboratory use.
(c) Foam	Mainly for small fires of burning petrol, oil, benzene and other immiscible solvents (Class B fires).	Dangerous for fires with an electrical risk.	Useful for flammable solvent stores.
(d) Tetrachloromethane	Mainly for small fires in the open	Produces poisonous fumes	Not suitable for college

Table 5.3 *(Continued)*

Type of extinguisher	Application	Disadvantages/ limitations	Notes
	air and solvent fires.	of carbonyl chloride. Violent explosions with sodium or potassium.	use. This extinguisher is now being phased out.
(e) B.C.F. (or 'Freon')	Can be used on virtually any fire (except Class D) and where there is an electrical risk. Leaves no residue.	Vapour slightly toxic and forms harmful pyrolysis products. Expensive.	Unsuitable for college use but widely employed in industry.
(f) Dry sand	Can be used on most small fires. The only readily available and effective first-aid treatment for **small** sodium, potassium and other Class D fires. Cheap and readily available.		A useful, general purpose extinguisher.
(g) Fire blanket	Virtually any small fire but especially Class B (burning liquids in a container) and for extinguishing burning clothing.		Should be available in all laboratories.

Assignment

List the most appropriate extinguishers for fighting Class A (carbonaceous substances), Class B (flammable liquids) and Class D (flammable metal) fires. Which extinguishers must *not* be used in cases where there is an electrical risk?

5.5
Use of fire extinguishers

Note: The fire fighting equipment provided in most laboratories is intended for *small* fires only. Portable extinguishers will cover two

Table 5.4 *Summary of fire extinguishers for different types of fire.*

Extinguisher	Class of fire † A	B	D ††	Suitability for fires in which an electrical hazard is also present
Water	✓✓	No	No	No
CO_2	✓	✓	No	✓
Foam	✓	✓✓		No
Vaporising liquids:				
CCl_4		✓	No	
BCF	✓	✓✓		✓
Dry powder	✓*	✓		✓
Dry sand	✓	✓	✓✓	✓
Fire blanket	✓	✓	✓	✓

† Class C (flammable gas) fires are treated by first turning off the gas supply if it is safe to do so and then extinguishing any remaining fire according to its class or type.

†† Specialised extinguishers are available and are used in industries where burning potassium or other metals is a particular hazard.

* General purpose powder extinguishers are available which are suitable for Class A (carbonaceous) fires.

square metres of fire at the most and are for 'first-aid' use only. Act immediately as a fire quickly grows. If anyone is near send them to report the outbreak and to assist in fighting the fire. The building must be evacuated and the fire brigade summoned immediately if a large outbreak occurs. Delay can be fatal as once a fire is out of control it can spread rapidly and cut off escape routes.

Always leave large fires or a situation in which a running solvent fire threatens gas cylinders or other solvent bottles to the fire brigade. Inform them immediately on arrival if inflammable solvents, gas cylinders (particularly hydrogen, ethyne (acetylene) and oxygen) or liquid oxygen are stored anywhere near the fire.

Remember: Asphyxiation by smoke and fumes is the principal cause of death in fires: people rarely die from the direct effect of the fire itself, i.e. by burns.

The following principles must always be obeyed when fire fighting:

1. Always take a position between the fire and the exit so your escape route cannot be cut off. Fire extinguishers should always be placed close to doors and other exits for this reason.
2. Do not continue to fight a fire if it is dangerous to do so or if there is a possibility that your escape route may be cut off by fire or smoke. A potentially fatal asphyxiating concentration of carbon dioxide can build up quickly if CO_2 extinguishers are operated in an enclosed space.

Close doors or windows behind you wherever possible if you have to withdraw.

The recommended procedures for operating the different types of fire extinguisher are:

(a) *Water extinguishers*
Direct the jet at the base of the flame and keep it moving across the fire. Attack a fire which is spreading vertically at its lowest point and follow the fire upwards. Concentrate the jet on any hot spots once the main fire is extinguished.

(b) *Carbon dioxide, dry powder and vaporising liquid extinguishers*
Fires produced by spilled liquids should be extinguished by directing the jet or discharge horn towards the near edge of the fire and with a rapid sweeping motion drive the fire towards the far edge until all the flames are extinguished (see fig. 5.8). Other types of fire may be extinguished by directing the jet directly at the burning material. The current should be switched off first if the fire is close to electrical equipment. The controlled discharge type of extinguisher may be turned off once the fire is out, but the fire should not be left unattended as reignition may occur. Vaporising liquid extinguishers should not be used in a confined space if there is a danger that the fumes may be inhaled.

(c) *Foam extinguishers*
If the burning liquid is in a container the jet should be directed at the inside edge of the vessel or at a vertical surface in order to break the jet and allow the foam to build up and spread across the surface of the liquid (see fig. 5.9). If this is not possible, the correct procedure is to stand well back (perhaps as far as 6 or 7 m) and to direct the jet as shown in fig. 5.10. With a gentle sweeping movement allow the foam to drop down and form a layer on the surface of the liquid. Do not aim the jet directly into the liquid as this will drive the foam under the surface

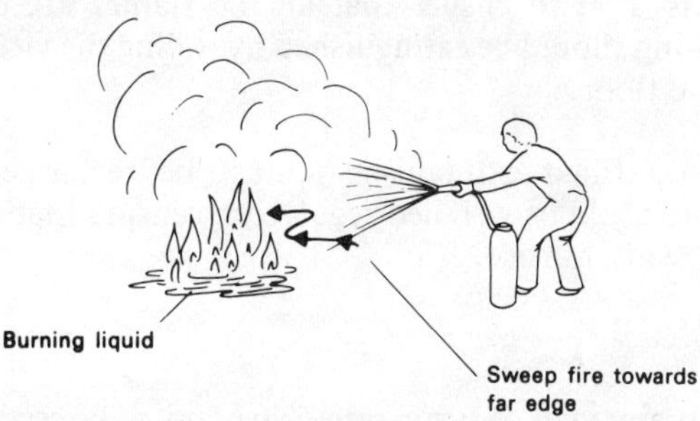

Fig.5.8 Use of carbon dioxide, dry powder and vaporising liquid extinguishers

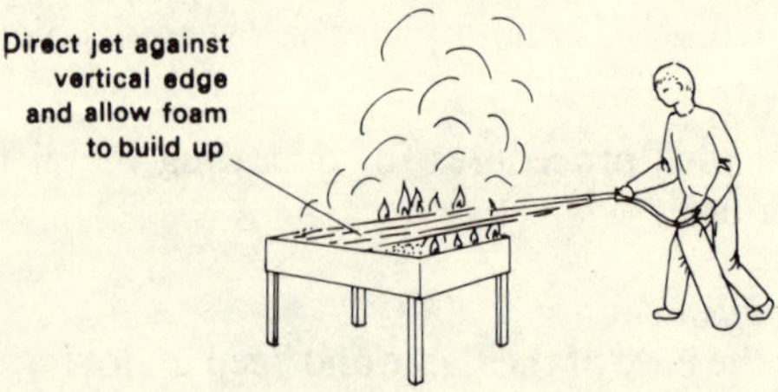

Fig.5.9 Use of a foam extinguisher

Fig.5.10 Use of a foam extinguisher

where it will be ineffective in extinguishing the fire and may spread the fire by splashing the burning liquid on to the surroundings.

(d) *Fire blankets*
A fire blanket may be used in conjunction with a carbon dioxide extinguisher, for example, for flammable liquid and other fires. The fire is first smothered with the blanket and the carbon dioxide extinguisher is used to ensure that all the flames are extinguished. Burning clothing should be extinguished by rolling the victim in the fire blanket on the floor.

Note: Partly filled extinguishers must be recharged after use. Extinguishers should be weighed regularly to ensure that they are fully charged and ready for use.

Assignment

Examine the main types of fire extinguisher and make certain you know how to operate them. Your college or firm may organise a practical session to give you experience in the use of fire fighting equipment.

5.6
Position of fire alarm buttons

5.7
Escape routes and assembly points

There is often considerable panic when a major fire breaks out, especially in a crowded building. People become confused and frequently behave irrationally when confronted by smoke and flames. A fire can spread quickly and it is essential to know the position of fire alarm buttons, of escape routes and of assembly points outside the building. There may not be time to enquire or look for directions once the fire has broken out and in any case smoke may obscure the emergency exit signs. The alarm should be sounded immediately to summon the fire brigade and to warn everyone to evacuate the building.

Do not wait to collect your coat or belongings as the delay could cost you your life. The lift must not be used as a means of escape as the lift shaft may quickly fill with fumes and the fire may cause a power failure. The stairs provide the safest escape route. One or two people on each floor should quickly check all the rooms, including offices, storerooms and toilets, on their way out to ensure that no one is left in the building and to close doors and windows wherever possible (see 5.8). Once outside the building everyone must report directly to the appropriate assembly point so that numbers can be checked. Do not wander off into town or go across to watch the arrival of the fire engines as anyone who is missing might be assumed to be trapped in the burning building and a life may be risked in a rescue attempt.

Assignment

Note the position at college and at work of the following:

1. the fire alarm buttons;
2. the emergency exits and escape routes;
3. the assembly points appropriate to the parts of the building in which you are working.

5.8
Localisation of fire and smoke

The methods by which a fire can spread were described in sub-section 5.2. It is important to close doors wherever possible if time permits when evacuating the building as this helps to localise the fire and limits the spread of smoke and fumes. Windows should also be closed before leaving to reduce the air supply to the flames. An open window can act as a flue by discharging smoke and hot gases to the outside air and drawing in a fresh supply of oxygen to increase the intensity of the fire.

The same effect is obtained if a door leading on to a nearby staircase is opened.

Assignment

Examine the fire regulations and notices at work and in college. Do they provide sufficient information about:

1. the location of escape routes, emergency exits and assembly points;
2. the importance of closing windows and doors if this is possible when evacuating the building?

Questions: 5 *Fire*

5.1 What is the fire triangle and what is it used for?
5.2 What is a fire?
5.3 Why do solids have to be heated to a fairly high temperature before they will burn?
5.4 What are the four classes of fire? Give *two* examples of each.
5.5 Name *five* of the commonest ignition sources and *three* of the commonest fuels in industrial fires.
5.6 What is the difference (if any) between the terms 'flammable' and 'inflammable'?
5.7 Explain what is meant by the following terms:
(a) flash point;
(b) ignition temperature;
(c) explosive limits.
5.8 What is a fire door? Name three of its functions.
5.9 Describe the methods by which a fire can spread in a multistorey building.
5.10 What is the principal cause of death in fires?
5.11 Why are cushions and furniture containing polyurethane foams dangerous if a fire breaks out?
5.12 Why are fire doors fitted with self closing hinges which open *towards* the exits from a building?
5.13 Why is it important not to wedge fire doors open?
5.14 Name *four* common types of fire extinguisher. Why are these extinguishers painted in characteristic colours?
5.15 Suggest suitable extinguishers for:
(a) burning wood and paper,
(b) burning solvents,
(c) a small amount of burning sodium.
How do they act? What extinguisher would you use for these fires if they were close to electrical equipment?
5.16 Why is it unsafe to use a soda–acid extinguisher on a smouldering transformer?
5.17 Why are asbestos fire blankets no longer used?

5.18 Why is it dangerous to use (a) vaporising liquid and (b) carbon dioxide extinguishers in an enclosed space?
5.19 Describe the procedure for using (a) a foam extinguisher and (b) a carbon dioxide extinguisher on a tray of burning solvent.
5.20 Why is it wrong to direct a jet of foam directly into a layer of burning oil?
5.21 Why are fire extinguishers kept close to doors and the exits from buildings?
5.22 What action should be taken if a fire is too large to be tackled with the equipment available?
5.23 Why is it important to tell the fire brigade if there are any compressed gas cylinders close to a fire?
5.24 What is the reason for reporting immediately to the appropriate assembly point after leaving a burning building?
5.25 Why is it important to know the position of fire alarm buttons, emergency exits and escape routes?
5.26 Why is it important to close doors or windows if possible when evacuating a burning building?
5.27 Why is it essential to recharge partly emptied fire extinguishers?

Section 6: *The expected learning outcome of this section is that the student should be able to recognise approved international warning signs*

Specific objectives: *The expected learning outcome is that the student:*
6.1 Recognises the international signs for radiation hazards.
6.2 Recognises the international sign for high voltage.
6.3 Recognises chemical and biological hazard signs.

Introduction

All warning signs need to be clearly visible. It is important that people should be immediately aware not only of the presence of a hazard, but also of its nature. Of course, this warning could be conveyed by a printed notice, but this does not always have the necessary impact and we all know that typed notices pinned to a board or printed instructions on a packet or container frequently go unread. In addition, the contents of the warning notice may not be immediately intelligible to the reader and may even be in a different language. There is now greater mobility in employment in Europe and throughout the world and the import and export of fuels, chemicals, drugs and raw materials is widespread and on a truly international scale. The use of hazard warnings which do not depend solely on language and the written word is therefore essential. These warnings must apply not just to the labelling of containers, but to marking out physical hazards, to indicate whether access to certain areas is forbidden, and to provide warning of an approaching danger etc.

One method is to use a colour code and to employ bright, strong colours to mark out hazards and to draw attention to escape routes, exits etc. The most powerful and distinctive colours are red, yellow and orange, and these are employed for the most urgent functions. Red is used for stop signal lights and danger warnings or for fire notices and equipment, while yellow and black diagonal stripes are used to make obstructions and other hazards more conspicuous. Colours are also employed to identify service pipes, conduits etc. and to signal the location of first-aid points. Green is used as the colour for safety and the British Standards Institution recommends that first-aid boxes should be painted in this colour for easy identification.

However, warning colours alone are not sufficient as not everyone will be aware of the colour code which applies and some may be colour blind. Hazard warnings are therefore indicated by symbols which are either painted in the warning colour or more clearly in black on a background of the relevant colour. Symbols have the advantage of being immediately recognisable and—as the message they convey is generally obvious—are not dependent on a knowledge of language. A number of international road signs come into this category. The international warning signs for radiation, high voltage, chemical and biological hazards are described in the following sections. The Hazchem code used on tankers, lorries or other road vehicles and on large containers of chemicals to indicate the proper action to take in case of an accident is summarised in Appendix VI.

6.1 International signs for radiation hazards

The principal sources of radiation hazard are radioactive substances or sources (see 4.1), X-ray tubes and lasers (see 4.4). Laser beams and ultra-violet, infra-red and microwave radiation are all examples of non-ionising radiation. The international warning symbols for ionising radiation and laser hazards are shown in figs. 6.1 and 6.2 respectively. The black trefoil sign on a yellow background should be placed on the door of all rooms or laboratories where X-rays are used or where more than trivial quantities of radioactive materials are present. It is not a danger sign and should not be labelled as such, but is intended as a

Fig.6.1 International sign for ionising radiation and/or radioactivity

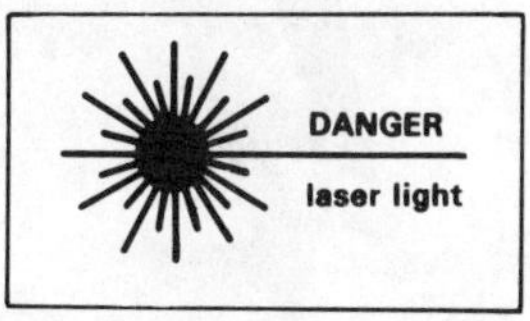

Fig.6.2 Warning sign for laser hazard

warning that ionising radiation may be present and that caution is needed. The sign should be surmounted by the word 'FIRE' in red if the room contains sufficient radioactive material to be a hazard to firemen in the event of a fire.

6.2
International sign for high voltage

The danger of electrocution increases with high voltage systems because a higher potential enables a larger current to pass through the skin. The electrical resistance of dry skin is about 10 000 ohms (see 1.7) so only a tiny current of less than 1 mA would be possible with a 6 V supply:

From Ohm's law:

$$\text{current} = \text{voltage}/\text{resistance} = 6/10\,000$$
$$= 0.6 \text{ mA.}$$

If however you were to touch the bare wire of a 5 kV supply while in contact with a metal pipe or a similar good conductor to earth, the current passing through the body would be almost a thousand times this value and would certainly be fatal. At very high voltages sparks are able to jump across small air gaps or across insulating materials and it is possible to be electrocuted when close to (but not touching) a high tension conductor.

A very high resistance (5–10 MΩ) is usually connected in series with the supply terminal in equipment which operates at a very high voltage but requires only a tiny current. This reduces the risk of electrocution by limiting the current that can be drawn from the supply. The generation of X-rays provides an additional hazard when high voltages are applied to evacuated gas discharge tubes. Biologically damaging radiation may be produced with potential differences greater than 5 kV and for this reason the Department of Education and Science has issued special regulations which apply to the use of voltages above this value in schools and colleges.

High voltages are required for the operation of X-ray machines,

Fig.6.3 Warning symbol for high voltage

lasers, electron beam deflection systems, electrostatic precipitators and many other pieces of equipment in medical, research and physical laboratories.

The high voltage warning symbol (fig. 6.3) should be placed on all such equipment close to the high tension connections or on the doors of rooms or laboratories where it is operated.

6.3
Chemical and biological warning signs

The EEC chemical hazard warning symbols for substances which are explosive, toxic, oxidising, corrosive, flammable and harmful or irritant are shown in fig. 6.5. Most of these symbols and especially those for explosive, corrosive and flammable substances are self explanatory, but nevertheless the sign should always be accompanied by the relevant printed hazard warning. These symbols are now widely used on labels for chemical containers (fig. 3.9). The symbol for labelling chemical containers of radioactive materials is shown in fig. 6.6).

The biohazard symbol (fig. 6.4) should be used to label all containers of potentially dangerous biological materials. The sign should be placed on the doors of all rooms or laboratories where work with pathogens is carried out or in which hazardous biological substances are stored as a warning to anyone entering to exercise care.

Assignment

Make certain you are able to recognise the international warning signs for radiation, laser light, high voltage, and for chemical and biological hazards.

Questions: 6 *International warning signs*

6.1 Why are symbols used as well as the printed word as a warning of the presence of a potential danger?

Fig.6.4 Warning symbol for biological hazard

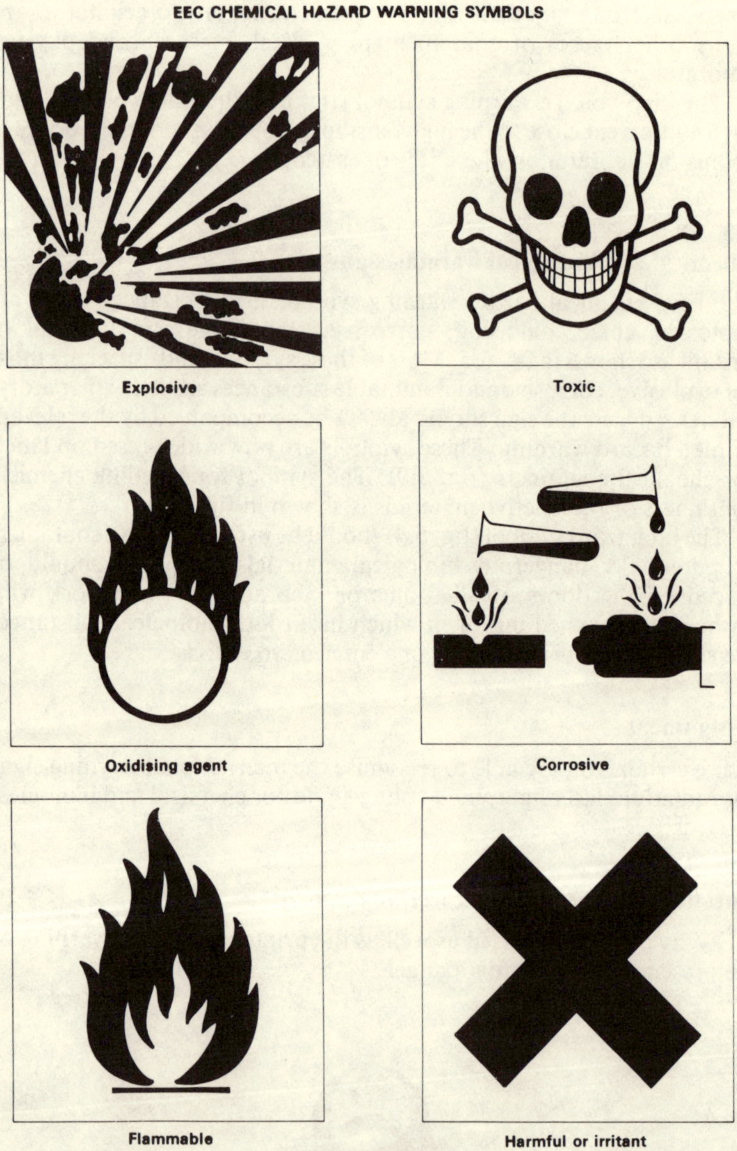

Fig.6.5 Warning symbols for chemical hazards

6.2 Give *three* examples of the use of colour codings to identify safety equipment and hazards.
6.3 Draw the warning symbols for radioactivity or ionising radiation,

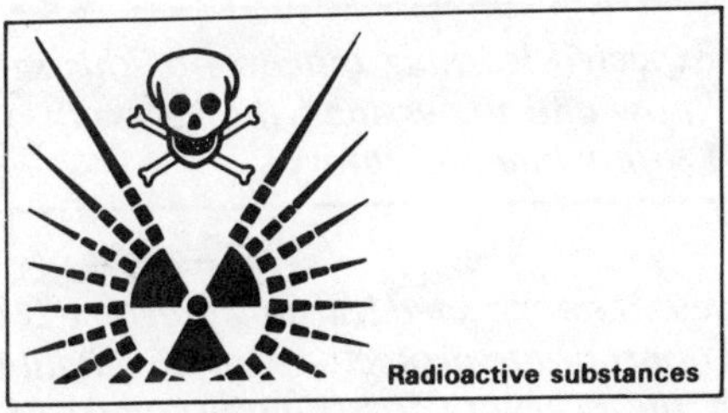

Fig. 6.6 Symbol for labelling chemical containers of radioactive materials

lasers and high voltage. Where are these symbols used and what is their function?
6.4 Draw the symbols for biological hazard and for the different types of chemical hazard. Where are these symbols used?
6.5 What is an oxidising agent? Give *two* examples and explain why such substances are hazardous.
6.6 What is the reason for fitting a high resistance in series with the supply terminal in some equipment which operates at high voltage?

Section 7: *The expected learning outcome of this section is that the student should know and understand the necessity of maintaining a personal code of safety in a laboratory*

Specific objectives: *The expected learning outcome is that the student:*

7.1 *Wears the appropriate protective clothing at all times.*
7.2 *Recognises the necessity for a mature code of behaviour in a laboratory.*
7.3 *Identifies the principles behind the establishment of a code as in 7.2*
7.4 *Explains the hazards from smoking, eating and drinking in a laboratory.*
7.5 *Explains the hazards of storing food or drink in a laboratory.*
7.6 *Lists a sequence of actions when going off duty or after handling dangerous substances, e.g. disposal of protective clothing, washing hands.*
7.7 *Recognises the correct methods of lifting heavy objects.*

Introduction

The chapters in the first half of this book described the principal hazards of working in a laboratory and the methods used for their control. These hazards stem from electricity; power tools and moving machinery; toxic, flammable, corrosive, carcinogenic and other harmful chemicals; cutting tools and damaged glassware; radioactivity, ultra-violet radiation and lasers; the risk of infection, fire and explosions. We saw too that the safety and health of anyone working in a laboratory depends not just on his or her own actions but on the actions of others. A moment's carelessness can injure the person working on the other side of the bench or in a different part of the laboratory. Many of the precautions against such accidents are obvious, but nevertheless are frequently ignored. For example, test tubes should never be pointed at other people, nor should bags or brief-cases be left on the floor where people can trip over them.

It may appear to be an exaggeration, but it is nevertheless true, that virtually all accidents are avoidable provided everyone working in the laboratory strictly observes a personal code of safety and obeys a few simple rules at all times. Teachers, chief technicians and senior laboratory staff should always set a good example. The aim is to instil the correct attitude to safe working and an awareness of the potential dangers of a particular operation and how these may be minimised. A general code of practice is incorporated within the laboratory rules listed in sub-section 7.2, but again it must be emphasised that lists of rules or regulations alone are not sufficient, no matter how prominently they are displayed. The important factor is that the regulations should be *obeyed* and that everyone should *understand* the reasons for observing them.

7.1
Protective clothing

The appropriate clothing for a chemical laboratory is a lab coat and a pair of safety spectacles or goggles. The lab coat should fit well and should be buttoned up correctly at all times. Cotton lab coats are more suitable than those made from nylon as they absorb more liquid and therefore offer more protection against spilled chemicals. An added advantage is that they do not generate sparks by static electricity which might ignite highly flammable organic solvents. Many people find it uncomfortable to wear safety goggles for any length of time, but this

Fig.7.1 Use of transfer chamber for manipulating micro-organisms

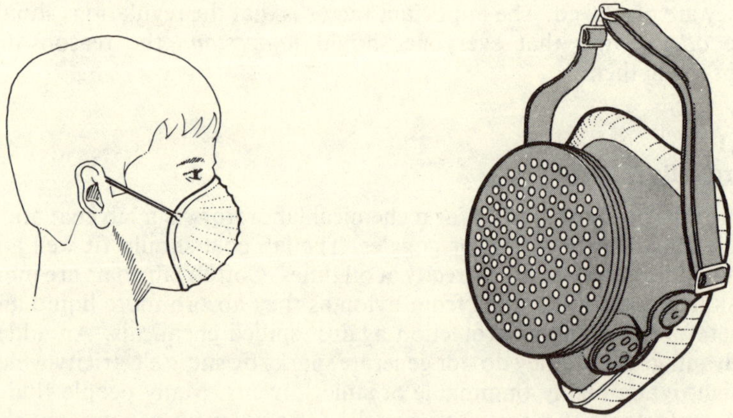

Fig.7.2 A dust mask

protection must be worn where strong acids, alkalis and other dangerous substances are used or where there is the slightest risk of splashes of chemicals or fragments of dust, grit, glass etc. getting into the eyes. Gloves should also be worn when transferring toxic, radioactive and carcinogenic compounds, irritants and corrosive liquids. The practice of wearing rubber gloves continually for laboratory work is not recommended as the hands become very moist and sweaty and skin infections or dermatitis may result. Rubber gloves also make it difficult to grip wet glass and serious accidents may result from dropped bottles or glassware.

Additional protection, such as a rubber apron and rubber wellingtons, is recommended for work with appreciable amounts of chromic acid, hydrofluoric acid and other highly corrosive liquids. A dust mask (see fig. 7.2) should be worn when transferring quantities of powders or grinding chemicals by hand. Gas masks or respirators are essential when working with highly toxic materials or when cleaning up large amounts of spilled harmful liquids. A respirator with oxygen bottle or cylinder is considerably safer than any gas mask in an emergency as it has its own oxygen supply. Gas masks can give a false sense of security and offer no protection in situations where the oxygen content of the air may be dangerously low, as for example after a fire, or where the gas mask absorbent is not suitable for the toxic gases or fumes which are present. The appropriate protective clothing for operating power tools was described in sub-sections 2.1–2.3 and the use of goggles with ultra-violet light discussed in sub-section 4.3.

7.2
The necessity for a mature code of behaviour in a laboratory

7.3
The principle behind the establishment of such a code

A laboratory is not the place for horseplay, games, practical jokes or other immature behaviour. Cuts from broken glass, burns from overturned bottles of acid and disfiguring scars from burning solvents are just three examples of the type of injury which may be suffered by the innocent victims of such foolishness. People should not run in laboratories nor should they throw books or other objects to each other which might knock over reagent bottles or damage equipment. The safety of everyone working in such a potentially dangerous environment can be ensured only if they behave in a mature and responsible manner and observe the following rules.

Rules/code of practice for behaviour in laboratories

1. Always wear proper protective clothing (e.g. lab coat and safety goggles or spectacles) at all times (see 2.1–2.3 and 7.1).
2. Make sure you know the positions of the main stopcocks or switches for turning off supplies of water, gas and electricity to the laboratory.
3. Make sure that you know the position of the first-aid kit, fire extinguishers, respirators and other safety equipment and that you know how to use them.
4. Never eat, drink or smoke in a laboratory (see 7.4 and 7.5).
5. Don't look into the mouth of a test tube or flask while you are heating it or adding the reagents. Never point test tubes at other people.
6. Check that all bunsen burners are out and that there are no naked flames before using inflammable solvents. Remember to warn everyone in the vicinity of the fire risk.
7. Report any breakage, faulty equipment and any other hazards immediately.
8. Wipe up any spilled chemicals immediately, especially corrosive acids or alkalis and mercury.
9. Do not run or play about in laboratories.
10. Do not store reagents and bottles of solvents in direct sunlight.
11. Put away any apparatus no longer required.
12. Do not sniff materials which may be toxic and never taste chemicals or eat seeds or parts of plants provided for biological studies.
13. Always use a fume cupboard for transferring highly toxic substances or for carrying out experiments which may produce harmful gases (see 3.1). Aerosols, e.g. of ninhydrin for the development of chromatograms, should be used in a fume cupboard.
14. Always label containers correctly with the full name and concentration of the contents (see 3.6).
15. Never try to stop or slow down a centrifuge with your hands. The speed at the outer edge may be greater than 150 km h^{-1} (100 m.p.h.).
16. Always wear a face shield when diluting strong acids and add the

acid in small amounts at a time with stirring to the water. Do not add water to the acid.

17. Always use mechanical fillers or safety bulbs when pipetting (see 9.8).
18. Do not charge accumulators close to naked flames (see 11.5).
19. Always wash your hands before leaving the laboratory (see 7.6).
20. Obey all safety rules at all times.

One's personal safety in a laboratory is determined to such a large extent by the behaviour of others that an excellent case may be made for excluding anyone who is not prepared to behave in a mature manner from the laboratory. The relevant provision of the Health and Safety at Work Act is discussed in sub-section 8.3.

Assignment

What are the codes of safety or rules which govern the behaviour of the people working in your laboratory? Do you consider them adequate and are they always obeyed? Suggest methods of making people more aware of the potential hazards and of encouraging them to obey the safety rules at all times.

7.4
The hazards from smoking, eating and drinking in a laboratory

7.5
The hazards of storing food or drink in a laboratory

The consumption of food or drink in any laboratory should be strictly forbidden as accidental contamination by chemicals or, in a biological laboratory, by bacteria can so easily occur. For the same reason food and drink should not be stored in the laboratory or in refrigerators which are also used for chemicals or biological specimens. The habit of smoking in a laboratory is also hazardous. It presents a serious fire risk when inflammable solvents are being used in other parts of the laboratory and—as well as being unpleasant to non-smokers—the presence of tobacco smoke can interfere with analytical work, purifications and other chemical and physical processes.

Some compounds which are comparatively harmless themselves can produce toxic substances when passed through the hot zone of a lighted cigarette or pipe. Examples of such substances include trichloromethane (chloroform), trichloroethylene and other chlorinated hydrocarbons which are partially oxidised by this treatment to yield the highly toxic gas known as (phosgene) carbonyl chloride, $COCl_2$, TLV 0.1 p.p.m. 0.4 mg m^{-3}. The products obtained by heating minute amounts of organic fluorine compounds, such as the polymer 'Teflon' (or p.t.f.e.—polytetrafluoroethylene), with tobacco can produce symptoms very similar to those of acute influenza within a few hours.

7.6 Sequence of actions when going off duty or after handling dangerous substances

Care should always be taken at the end of the day to ensure that the laboratory remains a safe place so that accidents are unlikely to occur during the night and at other times when no one is present. As a general rule equipment should not be left running, but if this is necessary particular attention should be paid to devising a 'fail safe' method of operation. Flooding is one of the commonest mishaps when apparatus is left on overnight. Surprisingly this can also be a cause of fires as the water seeps through floors and into ducts carrying electrical cables and causes short circuits. Allowance should be made for wide and sudden fluctuations in water pressure as factories close down for the night and water consumption drops from its daytime peak.

Laboratories in which there is a considerable fire risk should be fitted with smoke detectors and, where appropriate, automatic sprinkler systems. In an automatic sprinkler system water is carried to every part of the protected building in pipes suspended from the ceiling or roof. The water is held back by automatic valves until a fire outbreak occurs (see fig. 7.3). When the temperature rises above a predetermined value (such as 68 °C in many systems) a bulb of liquid ruptures or in some models a heat sensitive element fuses and allows the water to escape. An automatic alarm operates at the same time and the water sprinklers in the immediate vicinity of the fire either extinguish it before it has time to develop or contain it until the arrival of the fire brigade.

The following sequence of actions is recommended on going off duty:

1. Turn off equipment, bunsen burners, taps etc.

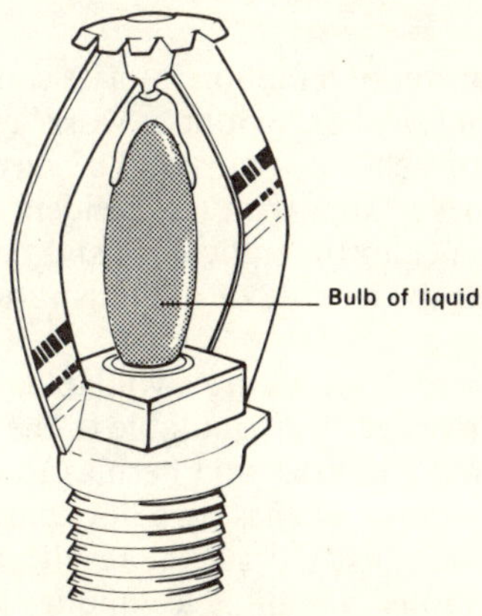

Fig.7.3 The automatic valve of a sprinkler system

2. Return reagent bottles to shelves and put away apparatus, chemicals and other materials. Radioactive sources (see 4.1) and poisons (see 3.2) must be locked away. Remember, the next people to come into the laboratory, e.g. the cleaners or caretaker, may not be aware of the hazards presented by such materials.
3. Place any other materials in suitable containers, label them and put them away safely.
4. In a biological laboratory check aquaria, animal cages and make certain the doors are secure. Ensure that the animals have sufficient food and water. Check that thermostats and oxygenaters for aquaria, incubators and other equipment in continuous operation have not been switched off.
5. Clear up any spilled chemicals etc. and wipe the top of the bench with a damp cloth.
6. Remove lab coat, safety goggles and other protective clothing and leave them in the laboratory or in your locker. Disposable gloves should be discarded and placed in a suitable receptacle for this purpose.
7. Wash your hands. The hands should always be washed when leaving the laboratory or after handling dangerous substances or material which carries a risk of infection (see 4.5).

Assignment

The procedure on going off duty described in this section was a general one. What modifications to this sequence of actions would you suggest for the laboratory where you work?

7.7
The correct method of lifting heavy objects

(a) *Introduction*

In any year up to a quarter of a million injuries caused by what may be described as 'wrong physical behaviour at work' are reported, while a large number of minor aches or pains and a considerable amount of physical discomfort never appear in the accident statistics. The total cost of the inconvenience to the individual and the loss in man hours and production added to the cost of administration and insurance is enormous.

Virtually all of these injuries are avoidable. Nearly a thousand injuries every week are directly attributable to the handling of objects and the majority of back injuries and hernias result from lifting and carrying. Surprisingly, most of these are not caused by attempting to lift excessively heavy weights, but are the result of adopting incorrect body positions and faulty handling techniques when moving comparatively minor loads. Most hernias and spinal injuries are the result of a cumulative strain over long periods.

A *hernia* is the protrusion of one of the internal organs through a gap in the walls of the cavity in which it is contained. The commonest type is the rupture of the abdominal wall by a loop of the intestines. This is generally caused by bending and lifting with the legs straight and the feet together. This compresses the abdominal contents and forces them against the walls where a rupture of the weak areas at which there is little muscular protection and support can occur. Lifting loads in this manner is also a major cause of spinal damage.

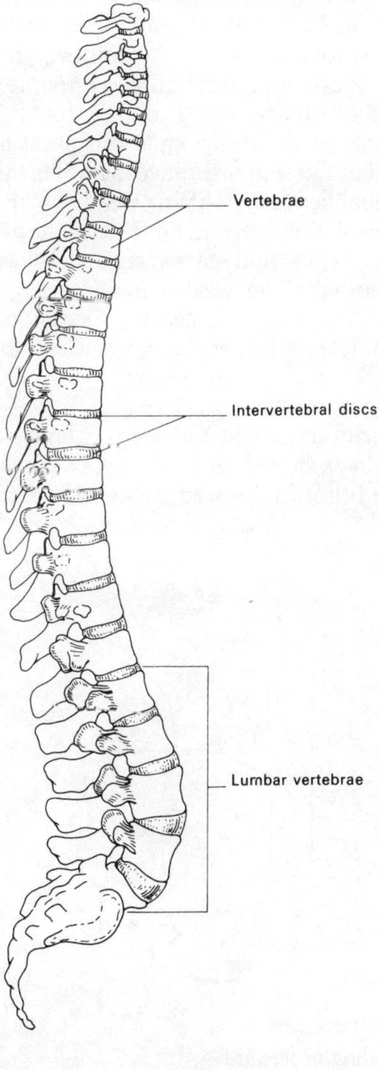

Fig.7.4 The spine

(b) *The spine*

The spine is a flexible structure consisting of thirty-three separate bones known as *vertebrae* (fig. 7.4). The nine vertebrae at the base are fused together and flexibility of movement is provided by the remaining twenty-four. Pads of fibrous tissue (known as *intervertebral discs*) between these movable bones act as shock absorbers. Each disc has a soft, gelatinous centre surrounded by a hard outer cover of cartilaginous tissue.

Provided the spine is upright any compressional force from lifting or carrying loads is distributed throughout the entire length of the column and heavy weights can be lifted without harm. But if the spine is bent, most of the stress is concentrated at the point at which bending occurs, particularly at the fourth and fifth lumbar vertebrae where movement is most free. Merely bending the spine to touch the toes exerts a compressional force of about 225 kg at this point in the average adult and every 25 kg of additional resistance increases this stress by as much as 150 kg. Continual bending or lifting weights with the spine curved in this position places all this stress on one side of the disc. This pinches the disc between the vertebrae and scars and gradually wears away the hard outer cover. Eventually the weakened disc may burst under compression, forcing out the soft centre. Occasionally this injury is produced immediately by heavy lifting while the spine is bent.

(c) *Correct lifting procedure*

To avoid the injuries described and the possibility of fibrositis and of torn muscles, the back should always be kept straight and the feet apart when lifting. The following procedure should be adopted:

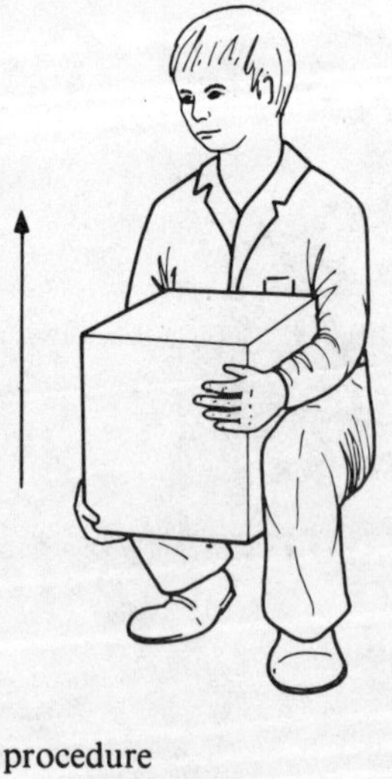

Fig.7.5 Correct lifting procedure

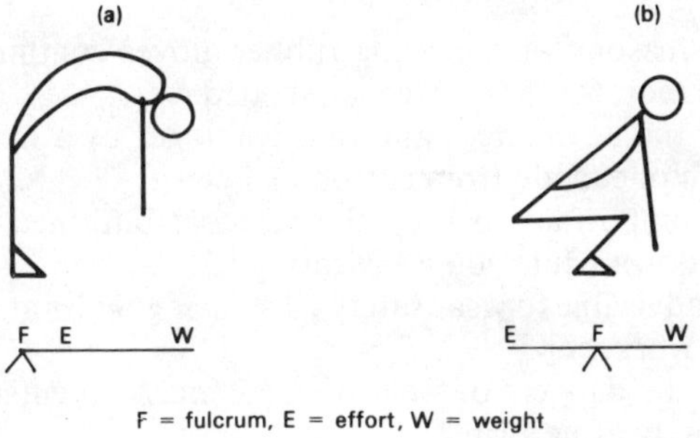

Fig.7.6 Mechanical advantage of (a) incorrect and (b) correct lifting procedure

1. Put one foot forward and to the side of the object to be moved. Always keep the feet about a hip breadth apart when lifting as this provides a larger base and improves balance.
2. Bend the knees to lower the body *vertically* to reach the object. *Always keep the back flat*. This prevents uneven stress being placed on the vertabrae as well as eliminating compression of the abdominal contents and strain of the back muscles.
3. Grip the object securely using the whole of the palms and not just the tips of the fingers. This will also reduce the chances of dropping the object.
4. Keep the arms close to the body with the arms as straight as possible so that the whole body can be employed to carry the load.
5. Keep the chin in. This locks the vertebrae and prevents injury to the cervical region (or neck) of the spine. It also makes it easier to breathe while lifting.
6. Straighten the legs without jerking and let the strong muscles of the leg do the lifting. Continue breathing normally.

The correct position for lifting objects is shown in fig. 7.5. This method gives a considerable mechanical advantage, as the diagrams in fig. 7.6(a) and (b) indicate. Where more than one person is lifting a heavy load, each should adopt the correct lifting procedure. They should all work in unison with just one person giving the orders to synchronise the movements.

Assignment

Practise the correct lifting procedure with a cardboard box or other suitable light loads until you have mastered it. Repeat the technique with groups of two or three people working in unison to lift a larger load.

Questions : 7 *Personal code of safety*

7.1 Give *two* reasons why wearing rubber gloves continuously while working in a laboratory is not recommended.
7.2 What are the advantages and disadvantages of a nylon lab coat compared with one made from cotton or linen?
7.3 Why is it important to keep the lab coat buttoned up with the sleeves rolled down while you are wearing it?
7.4 Why is it advisable to wear safety glasses or goggles at all times in a laboratory or workshop?
7.5 What are the dangers of wearing a gas mask to enter a chemical store immediately after a fire?
7.6 Why is it necessary to establish a mature code of behaviour to be followed by everyone working in a laboratory?
7.7 What are the hazards of eating, drinking or storing food in a laboratory?
7.8 Why is it inadvisable to smoke in a chemical laboratory?
7.9 Why is it advisable to wash one's hands after handling dangerous substances or when leaving a laboratory?
7.10 What is meant by 'fail-safe' equipment?
7.11 Why is it important to use thick-walled or reinforced rubber tubing and to wire it on to condensers which are left running overnight?
7.12 Why is it necessary to return reagent bottles to the shelves and chemicals to the stores at the end of a working day?
7.13 Why is it advisable to leave lab coats in the laboratory at lunch time and not to wear them in the canteen?
7.14 What are the major injuries which can result from attempting to lift a heavy load incorrectly?
7.15 What is the correct method of lifting a load?

B The law and technicians

Section 8: *The expected learning outcome of this section is that the student should be aware of the major provisions of national law and local regulations as they apply to technicians*

Specific objectives: *The expected learning outcome is that the student:*

8.1 *States local regulations concerning the disposal of stated dangerous substances.*

8.2 *States local regulations as applied at his/her place of employment for the handling of poisons, corrosive chemicals, radioactive sources and biologically hazardous substances.*

8.3 *States the main provisions of The Health and Safety at Work Act.*

Introduction

The hazards associated with toxic, flammable or corrosive chemicals and with radioactive sources and dangerous biological substances were described in earlier sections. All these materials are used in industry, hospitals, laboratories and in schools and colleges throughout the country. Immense environmental damage would result if quantities of these substances were to be disposed of into rivers, lakes or reservoirs and the storage of such materials in insecure or unsuitable containers close to schools or dwellings would present many hazards. It is not surprising therefore that laws have been enacted to control the purchase, transport, use, handling, storage and disposal of these materials. In addition to national law, many local authorities have issued their own regulations in an attempt to control and minimise potential hazards in their area. The major provisions of national law and local regulations as they apply to technicians are discussed in this section.

The following Acts apply—or are of interest—to technicians and to all people working in laboratories:

The Health and Safety at Work etc. Act 1974 The main provisions of this important Act are discussed in sub-section 8.3.
Carcinogenic Substances Regulations 1967.
Control of Pollution Act 1974.
Cruelty to Animals Act 1876 This Act controls (among other things) the use of vertebrates for experimental purposes.
Customs and Excise Act 1952 requires that laboratories should keep a record of supplies of ethanol, duty free spirit, industrial methylated

spirit and a number of other alcohols or alcohol–water mixtures.
Deposit of Poisonous Waste Act 1972 See sub-section 8.1.
Poisons Act 1972.
Radioactive Materials Act 1960.

A number of codes of practice or regulations concerning the use of dangerous substances, lasers, micro-organisms, radioactive materials etc. are discussed in the relevant sections of this book. These include:

The Use of Ionising Radiations in Educational Establishments, DES AM 2/76.
The Use of Lasers in Schools and Other Educational Establishments, DES AM 7/70.
Code of Practice for Chemical Laboratories, Royal Institute of Chemistry, 1976.
Laboratory Use of Dangerous Pathogens, DES AM 6/76.

8.1
Disposal of dangerous substances

The principal materials in this category are poisons, corrosive chemicals, flammable solvents, radioactive substances and materials which carry the risk of infection or can induce cancer, respiratory complaints, dermatitis and other injuries to health. The regulations governing the disposal of such substances vary from area to area and the technician should be acquainted with any regulations governing the disposal of the dangerous materials with which he is concerned. If he is in any doubt about the safe method to dispose of waste chemicals he should contact the firm's safety officer or an official from the local authority Department of Environmental Health for advice.

The Deposit of Poisonous Waste Act of 1972 penalises the depositing on land of poisonous, noxious (i.e. harmful) or polluting waste so as to give rise to an environmental hazard and makes offenders liable for any resultant damage. It also requires notice to be given in connection with the removal or deposit of waste. Where such waste cannot be made harmless before disposal the local authority, the river authority and/or the river purification board must be notified first and three clear days' notice given.

The following definitions are included in the Act:

'Land' here includes land covered with water and any part of the seashore.

'An environmental hazard' implies 'subjecting persons or animals to risk of death, injury or impairment of health, or contamination of water supplies'. The Act states that 'the degree of risk will be considered in relation to any measures taken to minimise the risk, and the likelihood of tampering by children or others'.

Protection of hazardous waste from tampering is essential and storage areas and sheds should be fenced and securely locked so that children, for example, cannot gain access.

The methods used for the disposal of *small* amounts of hazardous materials are described in the following paragraphs. Larger amounts present special difficulties and the method employed may be determined solely by local regulations and could demand the services of professional waste disposal firms. Accidents involving the spillage of large amounts of chemicals or the escape of explosive or poisonous substances require specialist advice. The Hazardous Materials Service at the Chemical Emergency Centre, Harwell, the Poisons Information Service and the TUC Institute of Occupational Health provide a telephone information service for safety and emergency information.

(a) *Disposal of radioactive materials*

Closed radioactive sources which are no longer required should be returned to the supplier. The Industrial Chemical Group at the Atomic Research Establishment, Harwell, should be contacted about the disposal of large amounts of radioactive waste.

(b) *Disposal of small amounts of waste chemicals*

Small quantities of liquid waste are usually washed down the sink with large amounts of water, but flammable liquids, carcinogens and highly toxic substances should never be discarded in this way nor should they be buried. The vapour of most volatile flammable liquids, such as ethoxyethane ('ether'), petrol, petroleum ether and propanone (acetone), is denser than air and tends to collect in drains or waste ducts. This can spread fire hazards to other parts of a building and explosive mixtures may be produced with air. Flammable liquids should be stored in labelled bottles for distillation and re-use or, if they cannot be recovered in this way, they may be disposed of by burning outside the laboratory or kept in a locked store (see 3.4) for collection by an industrial waste disposal firm. One useful method of disposing of accumulated supplies of flammable solvents is to burn them in a flat metal tray in the open air well away from any buildings. The resulting fire provides a valuable training exercise in the use of fire extinguishers (see 5.5). Most fire brigades are willing to send a representative to a college to demonstrate and supervise the use of firefighting equipment.

Unless the amounts are excessive or the material is particularly hazardous, waste gases and smoke are disposed of into the atmosphere via an effective hood constructed of noncombustible and corrosion resistant materials fitted to an efficient fume extraction system. Fume cupboard fans should be effectively protected and earthed.

Solid waste may be placed in refuse bins provided it is both inactive and insoluble. Glass waste should first be packed in special containers. Reactive substances must be converted into less reactive materials before disposal. Sodium or potassium may be destroyed by adding them in small amounts at a time to a large volume of anhydrous industrial (methylated) spirit. (*Care: safety goggles must be worn.*) Calcium dicarbide is disposed of by first adding it to water in small quantities at a time:

$$\underset{\text{calcium dicarbide}}{CaC_2} + 2H_2O \longrightarrow Ca(OH)_2 + \underset{\text{ethyne}}{C_2H_2}$$

This reaction must be carried out in the open air and not in the laboratory owing to the high flammability and explosive properties of the gaseous ethyne (acetylene) which is evolved. Acids should be neutralised before disposal, otherwise corrosion and leakage of metal drums used for subsequent removal may result.

(c) *Bacteriological cultures*
Bacteriological cultures should be soaked in aqueous 3% 'Lysol' or autoclaved before disposal to destroy pathogens. Disposable Petri dishes should be incinerated unopened.

Assignment

What are the methods used to dispose of dangerous substances at work and in the college laboratories? List the statutes and local regulations which apply to these activities.

8.2
Regulations for handling dangerous substances

Many of the recommendations from the regulations and codes of practice which govern the handling of poisons, corrosive chemicals, radioactive sources and biologically hazardous substances have been incorporated in the relevant procedures described in earlier sections. For example, the methods of minimising the dangers of infection and disease in medical or biological laboratories were listed in sub-section 4.5. Similarly, the procedures for working with carcinogens, radioactive sources and lasers were described in sub-sections 3.10, 4.1 and 4.4 respectively. The regulations issued by most local authorities to govern the handling of the dangerous materials closely follow these codes of practice and recommendations.

Assignment

State the local regulations which apply at your place of employment for the handling of poisons, corrosive chemicals, radioactive sources and biologically hazardous substances.

8.3
The Health and Safety at Work Act 1974

The Health and Safety at Work Act received the Royal Assent in July 1974 and came into force on 1 April 1975. The Act supplanted all or part of thirty-one other Acts and thereby simplified, strengthened and unified the legislation protecting people at work. It applies to virtually every person in the workforce and brought legal protection to over five million people for the first time. Among those to whom such legislation was extended are workers in the medical and dental services; central and local government employees; teachers in schools, colleges and universities and people engaged in research and development. The four main objectives of the Act are:

1. To secure the health, safety and welfare of persons at work.
2. To protect persons, other than those at work, against risks to their health or safety arising out of or in connection with the activities of persons at work.
3. To control the keeping and use of explosives or of highly inflammable or otherwise dangerous substances and to prevent people acquiring, possessing or illegally using such substances.
4. To control the emission of noxious or offensive substances into the atmosphere.

In order to achieve these aims the Act places a number of obligations on the employer and on the employee. For example, the employer is obliged to provide:

1. A healthy and safe working environment with adequate amenities.
2. Safe plant, machinery and equipment which has to be maintained in good order.
3. Adequate instruction and training for employees.
4. Adequate supervision by competent personnel.
5. Information to employees to ensure their health and safety at work.
6. A written safety policy.

The basic responsibilities of the employees are:

1. To act in the course of their employment with due care for the health and safety of themselves, other workers and the general public. The worker is required by law to use the safety devices and protective equipment provided and not as a matter of personal choice. Any worker who does not behave in a mature and responsible manner (see 7.2) or who—after previous warnings—continues to work in such a way as to place the safety of others at risk should be excluded from a laboratory.
2. To report any conditions they consider dangerous and to suggest ways in which the work could be made safer. Any hazards should be reported immediately and preferably in writing. It is advisable to keep a copy of this report and of the date when it was submitted. Examples of

such hazards which should be reported so that the employer will be able to rectify them before anyone is injured are electrical faults, damaged cylinder valves or gauges, gas leaks and faulty fume cupboards. If for any reason the equipment cannot be made safe, there is no alternative but to cease to use it.

It is also illegal for any employer to levy a charge on an employee for any safety equipment, goggles, protective clothing etc. provided to meet the legal requirements of the Act.

The Act allows a recognised trade union to appoint a safety representative and lists his functions and duties. For example, such a representative is required to investigate potential hazards, to attend meetings of safety committees and to inspect the workplace at three-monthly intervals and carry out inspections after an accident, dangerous occurrence or notifiable industrial disease.

The appointment of a safety officer is essential in any college, school or reasonably sized laboratory or firm. His (or her) main duties will be advisory and are concerned with the attainment of a safe working environment. The safety officer should be familiar with the type of work carried out in the laboratories and be fully aware of the hazards involved so that he can pass this information on to those operating the hazardous processes or working with potentially dangerous materials. He should also visit all the laboratories, stores, workshops, production lines etc. for which he is concerned at frequent intervals for informal consultations with individual workers. His advice on training, the selection of protective equipment, the preparation of safety instructions, the reporting and investigation of accidents, the strict observance of safety procedures etc. will then be felt to have more relevance and will carry greater weight. A safety officer has to work with the management to obtain detailed information about the properties and potential hazards of any new materials or processes and then make sure that those involved are aware of these hazards and of the safety and emergency procedures to be followed. It should be remembered that safety is not solely the responsibility of the safety officer: everyone is involved—from the manager or director to the most junior laboratory technician. Of course, it helps if the man or woman appointed as safety officer is a person of integrity who carries out the job conscientiously and commands the respect of all, but it in no way reduces the need for the observance of correct safety practices by all other laboratory workers at all times.

Assignment

Read the statement of safety policy issued by your employer to meet the requirements of the Health and Safety at Work Act. How often does the firm's safety committee meet and who are its members?

Questions: 8 *Law*

8.1 State the local regulations concerning the disposal of dangerous substances.
8.2 Why is it unsafe to pour ether, propanone (acetone) and other inflammable solvents down the sink?
8.3 Why is it advisable to neutralise acids before disposing of them?
8.4 What are the four main objectives of the Health and Safety at Work Act?
8.5 What obligations does this Act confer on (a) the employer and (b) the employee?
8.6 Why is it important to report any electrical faults or other hazards which develop at work? Why is it advisable to make such reports in writing and to keep a copy?
8.7 Does the Health and Safety at Work Act benefit members of the public?

C Laboratory practice and equipment handling

Section 9: *The expected learning outcome of this section is that the student should know the simple techniques of reading instruments and be able to state their limitations and accuracy*

Specific objectives: *The expected learning outcome is that the student:*

9.1 Uses a vernier scale to measure a distance.
9.2 Uses a micrometer to measure the thickness of a small object.
9.3 States the accuracy of measurements made with a vernier and a micrometer.
9.4 Uses a variety of modern balances to determine the mass of a quantity of a solid chemical.
9.5 States the accuracy of the mass determined in 9.4.
9.6 Checks one balance against another and states whether they need servicing or not.
9.7 Uses a burette to deliver a stated volume of a liquid.
9.8 Uses a pipette to deliver a stated volume of a liquid.
9.9 States the accuracy of the volumes delivered in 9.7 and 9.8.
9.10 Uses a multimeter to measure d.c. and a.c. potential and test for continuity.
9.11 Sets up and demonstrates the use of optical microscopes employing both transmitted and reflected light.
9.12 Demonstrates a correct procedure for the cleaning of microscopes.
9.13 States that standards for physical, chemical and biological quantities exist.
9.14 Demonstrates how a range of the more simple standards can be used to standardise instruments.
9.15 Constructs a schedule for the periodic testing of common laboratory instruments against such standards.
9.16 States the purpose of the BSI.
9.17 States the titles of the BS publications that apply to their own field of work.

Introduction

Chemistry, physics and biology are all exact sciences which aim to quantify their experimental observations wherever possible. An important part of laboratory work in any of these subjects is the use of instruments for the accurate measurement of such physical quantities as length, mass, volume, pH, time and electrical potential. Every technician should be so familiar with the use of such instruments that he or she can be confident in the reliability, accuracy and reproducibility of any measurements made, as well as being aware of their degree of accuracy. The purpose of this chapter is to help the reader to achieve these aims and to describe the standards available to check the accuracy of measurements made of these important physical, chemical and biological quantities.

(a) *Accuracy and precision*
It is important to distinguish between the related but separate terms: *accuracy* and *precision*. *Accuracy* may be defined as the closeness of agreement between the true value of the quantity which is being measured and the experimental value, while *precision* is the closeness of agreement obtained in a series of replicate (i.e. repeated) experimental values of the measurement. A precise measurement of a property is not necessarily an accurate one. Error is defined as the difference between the measured value of a property and its true value.

It is not always possible to express the reliability of a measurement in terms of its accuracy as the true value of the property may not be known. In many cases (e.g. specific heat, density and refractive index) only an accepted value is available. The accepted value of a property is its most probable value derived from repeated, careful measurement.

An indication of the precision of a measurement, such as the volume of acid required in a titration, may be found by repeating the measurement several times and recording it to the practical limit to which the instrument can be read, e.g.

$$V/\text{cm}^3 = (1)\ 26.20,\ (2)\ 26.15,\ (3)\ 26.15.$$

The consistent value, i.e. 26.15 cm^3, or the mean of the most consistent values is normally used. The precision of an experimental measurement depends on a number of factors, of which the most important are the nature and type of the instrument used to measure it and the skill of the operator. Error in the instrument may be eliminated by calibration against a suitable standard (see 9.13–9.15). This calibration determines (or checks) the limiting precision of the instrument.

9.1
The vernier

A vernier scale provides a convenient method of measuring distances to a far greater precision (and accuracy) than is possible with a metre rule or simple measuring tape. The human eye is unable to detect size differences of less than 0.1 mm from a rule without the aid of a lens or of some other magnifying principle. Divisions of this size would be too close together on a scale to be easily distinguished and methods have been found to magnify the movement of the measuring face so that larger divisions may be used. The two commonest principles by which magnification is achieved are the *vernier* and the *micrometer* (see 9.2).

The vernier scale depends on the fact that the eye can easily detect which of the fine divisions on two neighbouring scales are in line with one another. Magnification is provided by the slight difference in size of the division of the two scales: those on the sliding scale are usually shorter than those on the fixed scale (see fig. 9.1). The smallest distance to which the vernier can be read is equal to the difference in size of the divisions of the two scales.

The vernier shown in fig. 9.1 can be read to one hundreth of a centimetre. i.e. to 0.1 mm. The main scale is divided into 1 mm divisions, while the divisions on the vernier are 0.1 mm less or 1.0 − 0.1 = 0.9 mm. Ten divisions on the vernier scale are required to measure the 1 mm divisions on the main scale in 0.1 mm units, so the vernier scale is 10 × 0.9 = 9 mm in length. If the vernier scale is moved 0.1 mm from zero, i.e. from the position where the zero marks of each scale are in line, the first division on the vernier scale will be in line with the next division of the main scale. Further 0.1 mm displacements of the sliding vernier scale bring the second, third and subsequent divisions on the vernier in line with divisions on the main scale (fig. 9.2). An example of this principle to determine a vernier reading is shown in fig. 9.1.

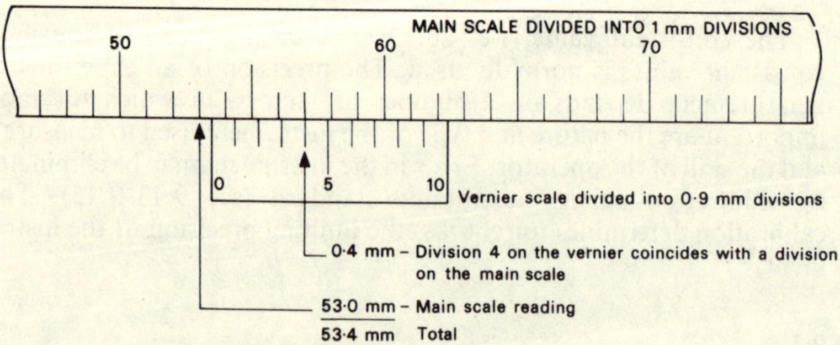

Fig.9.1 Vernier

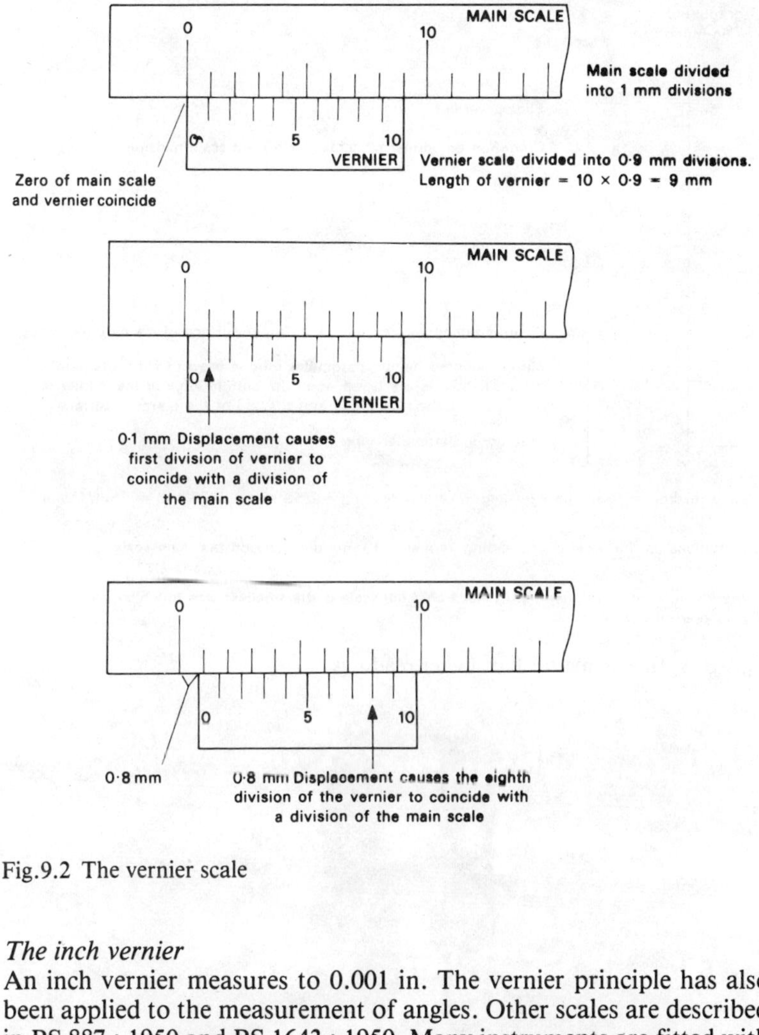

Fig.9.2 The vernier scale

The inch vernier

An inch vernier measures to 0.001 in. The vernier principle has also been applied to the measurement of angles. Other scales are described in BS 887 : 1950 and BS 1643 : 1950. Many instruments are fitted with verniers to improve their reading accuracy. Examples include the micrometer (see 9.2), the analytical balance (see 9.4), the barometer, (fig. 9.3) the mechanical stage of a research microscope (see 9.11), the vernier microscope (fig. 9.4) and slide callipers (fig. 9.5).

Vernier callipers

Vernier callipers (or slide callipers) are used for the accurate measurement of the internal or external dimensions of small objects. They consist of a steel scale with a fixed jaw at one end. The object to be measured is placed between a fixed jaw and a sliding jaw which is fitted

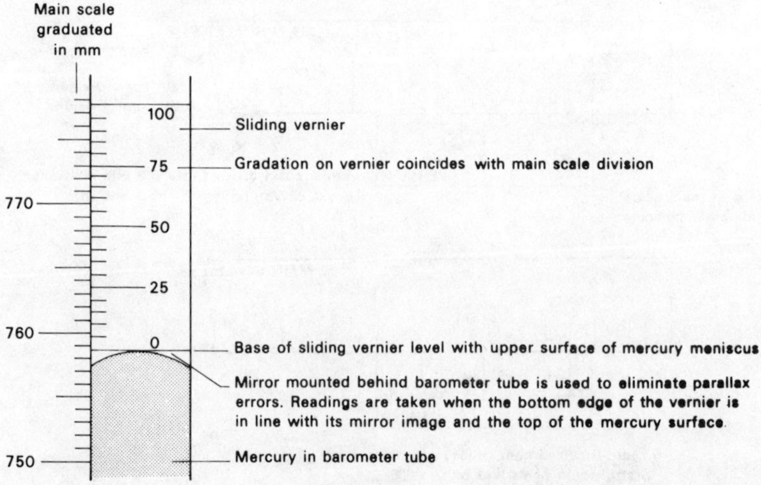

Pressure reading = main scale reading + vernier reading = 758 mm + 0·75 mm = 758·75 mm

20 gradations on the scale of the sliding vernier = 19 mm divisions on the main scale

Difference in length between the divisions on each scale is the smallest size to which vernier may be read = 0·05 mm

Fig.9.3 Vernier scale for barometer readings

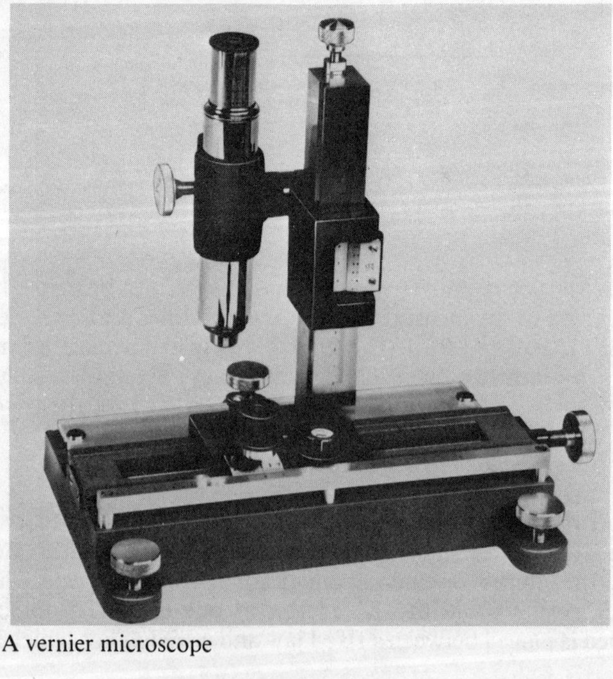

Fig.9.4 A vernier microscope

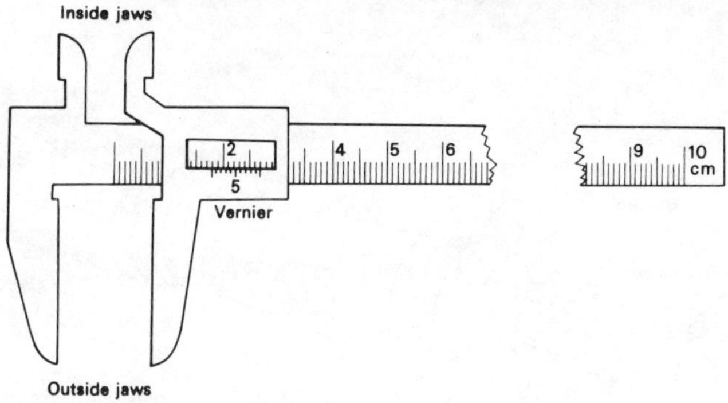

Fig.9.5 Vernier (slide) callipers

with the vernier and moves along the steel scale. Slide callipers are usually fitted with 'inside jaws' to measure the internal diameters of pipes and tubes, for example, as well as 'outside jaws' to measure external diameters or thicknesses.

Assignment

Measure the internal diameter and external diameter or thickness of a number of convenient objects using vernier callipers.

Question

What are the vernier readings shown in fig. 9.6(a) and (b)?

9.2 The micrometer

The principal parts of a micrometer screw gauge are shown in fig. 9.7.

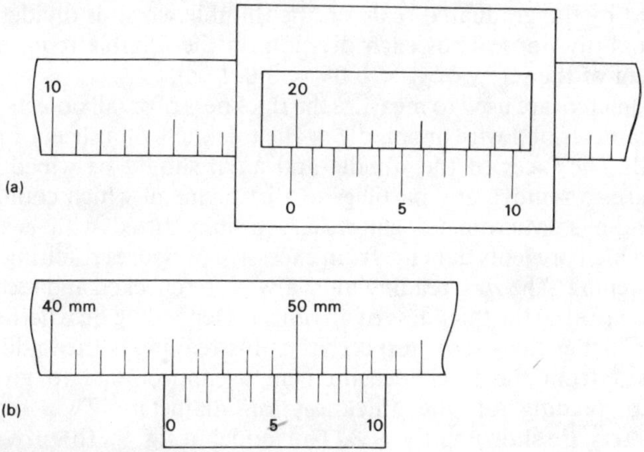

Fig.9.6 Vernier readings

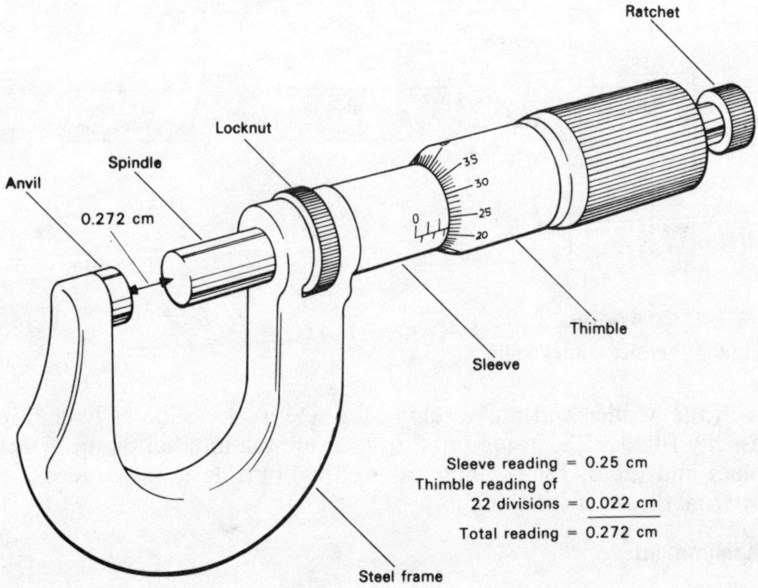

Fig.9.7 The micrometer screw gauge

Magnification for precise measurement is provided by a screwed spindle fitted with a graduated thimble. The screwed portion of the spindle is totally enclosed to protect it from damage and the sleeve is graduated in 0.5 mm divisions. If the pitch of the screw is 0.5 mm, the spindle will move through a distance of 0.5 mm or 0.05 cm (i.e. one division) for each complete turn of the thimble. Fractions of a turn are measured by the graduated scale on the thimble which is divided into fifty equal divisions. Thus each division on the thimble represents a movement of the screw of $\frac{1}{50} \times 0.05 = 0.001$ cm.

Micrometers are used to measure the thickness of small objects, such as the diameter of a wire or spindle or the thickness of a sheet of paper or metal. The faces of the spindle and anvil should be wiped clean before use to remove any particles of dirt or metal which could give false readings. Micrometer gauges are usually fitted with a spring ratchet which prevents damage from excessive pressures resulting from overtightening. The zero reading must always be checked and recorded before measuring the thickness of an object. Depending on whether it is lower or higher than zero, respectively, this reading is then added or subtracted from the final measurement as appropriate to give the corrected reading of the thickness or diameter. Two modern micrometers are shown in fig. 9.8. The model in fig. 9.8(b) provides a direct, digital readout.

(a)

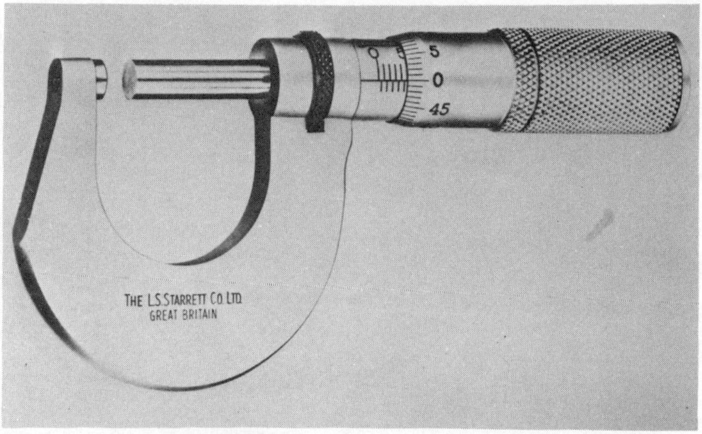

(b)

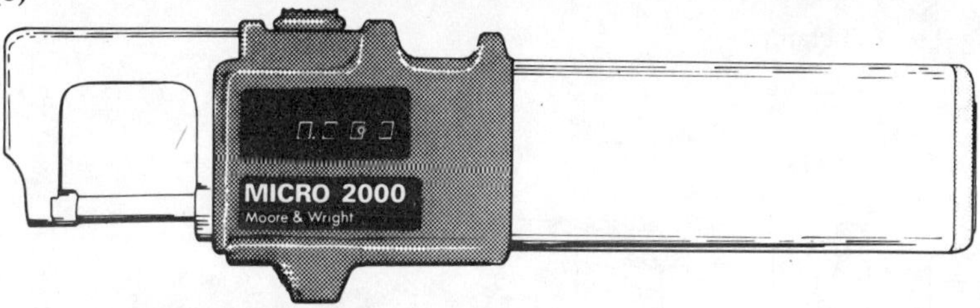

Fig. 9.8 (a) A micrometer screw gauge (b) A digital micrometer

Assignment

(1) Examine a G-clamp (fig. 9.9) and describe how it could be used to demonstrate the principle of the micrometer screw gauge. (2) Use a micrometer screw gauge to measure the diameter of a metal rod or of a piece of wire at a number of points in different directions along its length. Note the zero reading each time and apply the necessary correction to the measurements. Discuss the extent to which the readings reflect the precision of the micrometer gauge or the lack of uniformity in the circular section of the rod or wire.

Question

(a) What are the micrometer readings shown in fig. 9.10(a)–(c)?

9.3
Accuracy of measurements made with a vernier and a micrometer

The smallest distance to which a vernier may be read is equal to the dif-

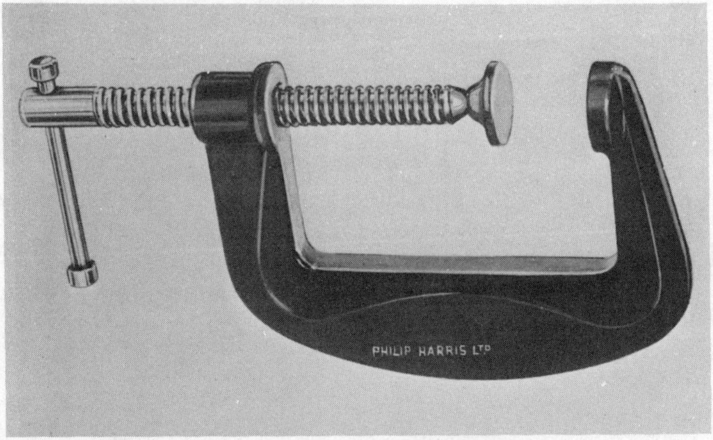

Fig.9.9 A G-clamp

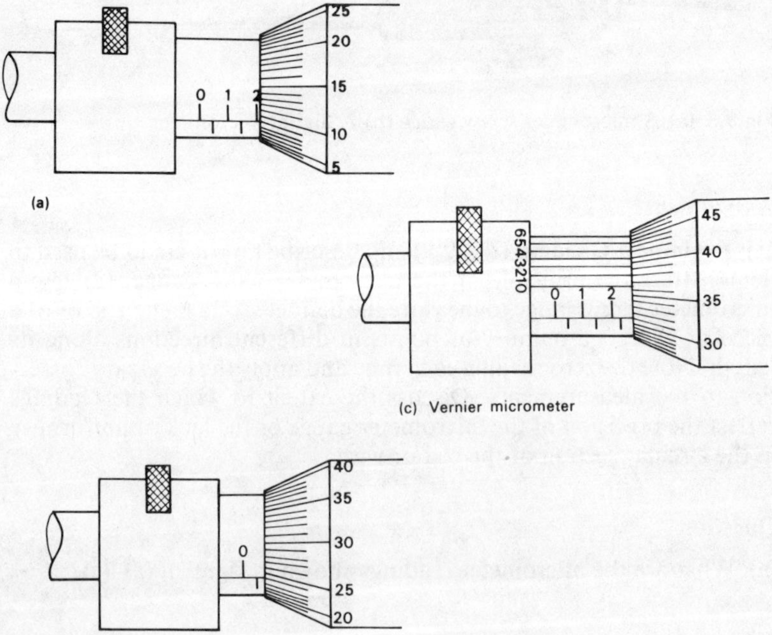

Fig.9.10 Micrometer readings

ference between the lengths of the divisions of the main scale and the vernier scale (see 9.1). The precision and—for a skilled operator using a calibrated instrument in correct working order—the accuracy of metric and inch verniers similar to those described in sub-section 9.1 are 0.002 cm and 0.001 in respectively.

The precision of a micrometer depends largely on the size of thimble and barrel as this determines the number of parts into which each 0.5 mm unit can be subdivided. In an accurate micrometer these units are divided into fifty equal parts, each representing 0.01 mm. The vernier principle may also be applied to micrometers to subdivide the thimble division into 10. The vernier scale is engraved round the barrel of the micrometer and make it possible to read the instrument to 0.0002 cm or 0.0001 in. High precision micrometers such as these are used mainly for testing gauges.

It is important when assessing the accuracy of a measuring instrument to distinguish between the precision to which the instrument may be read (i.e. its reading accuracy) and the accuracy to which the quantity is measured (i.e. the measuring accuracy). Sometimes the precision to which the instrument can be read can be increased by estimating fractions of a division, but the effect of this is limited by the accuracy to which the instrument can be set. The setting accuracy of a micrometer, for example, is about ± 0.002 mm. The measuring accuracy of an instrument is thus dependent on both the reading accuracy and the setting accuracy and, to a large extent, on the skill and experience of the operator. The accuracy to which measurements may be made consistently, based on a length of 25 mm, are listed in table 9.1.

Table 9.1 *Accuracy of vernier and micrometer*

Range	Measuring instrument	Tolerance/ measuring accuracy
15–100 cm	Steel rule	± 0.25 mm
0–300 mm	Vernier callipers	± 0.02 mm
0– 12 in		± 0.001 in
0– 25 mm	Micrometer	± 0.007 mm
	Micrometer set to gauge blocks (see table 9.5)	± 0.005 mm

Assignment

Refer to the manufacturer's literature to determine the measuring accuracy of the vernier callipers and micrometer gauges available in the laboratory.

9.4
Balances and their use

Most modern balances are of the single pan type in which weights are added to or removed from the beam by turning the knobs at the front of the instrument. In electronic balances the depression of the scale when the object is placed on it is proportional to the mass of the object being weighed. This depression is measured by a change in resistance of a strain gauge linked to the beam. The mass of the object is then indicated on a moving scale or displayed digitally. Two typical modern balances are shown in figs. 9.11 and 9.12. Construction varies from one manufacturer to another but—with slight differences in operating instructions from balance to balance—the general principles and weighing procedure are essentially the same.

A balance should be placed on a heavy weighing table (a stone slab is ideal) in a room which is free of vibration and draughts. The instrument

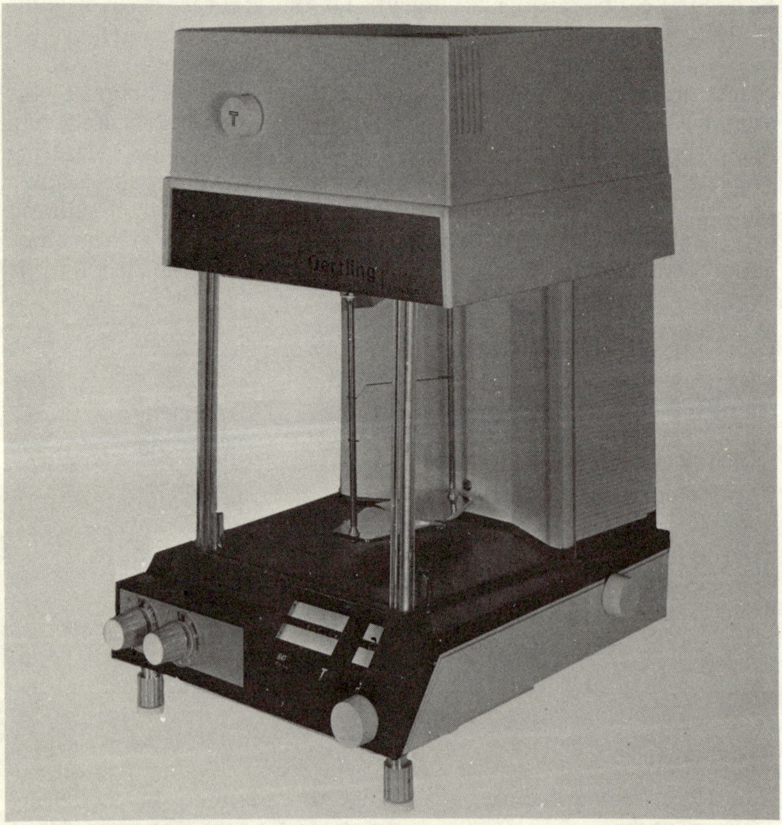

Fig.9.11 An analytical balance

Fig.9.12 An electronic top pan balance

should be levelled by turning the two foot screws at the front of the balance until the spirit level bubble is centred. Solids or liquids should always be weighed on a watch glass or in a beaker or glass stoppered weighing bottle (see fig. 9.13) and their mass found by subtracting the weight of the empty container from the combined weight of the container and substance. Powdered chemicals must not be weighed out directly on the scale pan nor on pieces of paper. Many balances have a tare facility which enables the instrument to be zeroed with the empty container on the scale pan so that the weight of the substance (i.e. the contents of the container) may be read directly on the scale.

Weighing procedure

The general procedure for determining the mass of a quantity of a solid chemical using an analytical balance is:

1. Set all the weight knobs to zero and check that the balance is level and that the scale pans are clean.
2. Turn the release knob and wait until the image is stationary before adjusting the zero of the instrument.
3. Arrest the balance by turning back the release knob and place the object to be weighed (in this case the weighing bottle or watch glass, i.e.

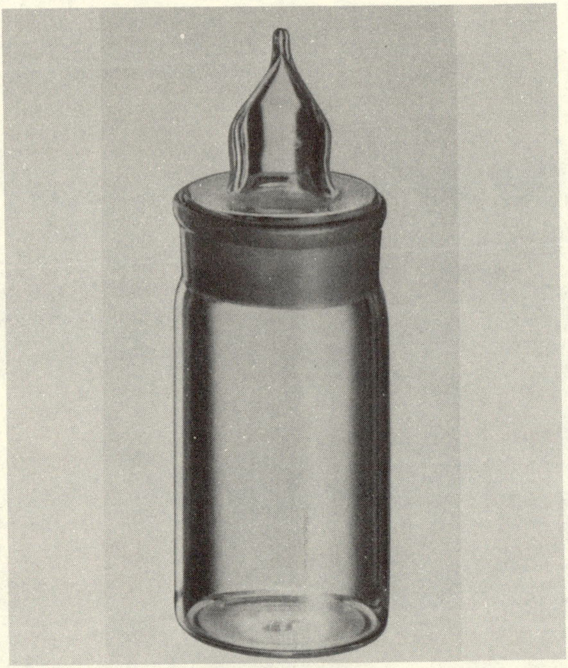

Fig.9.13 A glass stoppered weighing bottle

the container for the solid chemical) *gently* in the centre of the pan. Weights should never be placed on the pan (or added to or removed from the beam) when the beam is free on a non-electronic balance as this can damage the knife edges or bearings.

4. Close the slide or balance door to eliminate the effect of draughts.
5. Turn the weight knobs at the front of the instrument to obtain an approximate value of the mass of the object. Some balances have a facility for preweighing which gives the approximate weight immediately so that the weight knobs may then be set at once to this reading. If the balance does not have a preweigh facility the approximate mass is found by turning the weight knobs to progressively increase the mass added first in 5 or 10 g steps and then in 1 g steps. The balance is arrested for each addition.
6. Release the balance and turn the 0.1 g weight knob until a reading is obtained on the scale.
7. Wait until the image is stationary and then read off the precise weight of the object. The reading for grams and usually the first decimal place are taken directly from the weight knobs, while the remaining decimal places are taken from the projected scale. Occasionally a fourth decimal place may be read from the micrometer

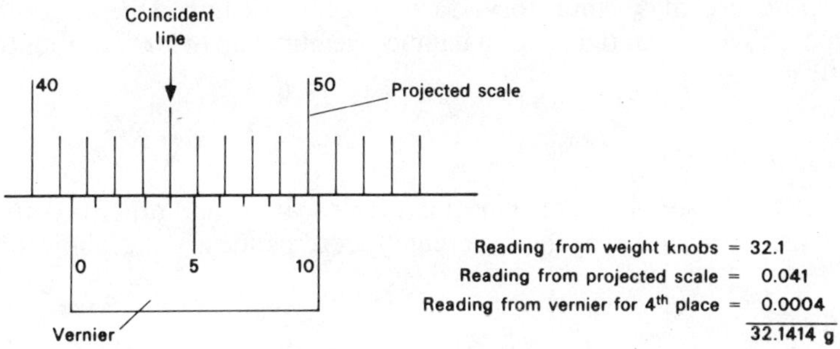

Fig.9.14 Use of vernier scale on an analytical balance to give fourth decimal place

or vernier scale on the instrument (fig. 9.14). The weight should be recorded immediately in a permanent notebook and not written on a piece of filter paper (see 10.1).

8. Arrest the balance. Add the required sample weight to that of the container and set the weight knobs to this value. Open the balance and add the sample to the container from a spatula. In most cases it is sufficient to add the sample until the reading comes on to the scale and then to record the exact weight. If, however, a certain, precise weight of the sample is required, the balance should be opened and released. Then, taking care not to touch the container, scale pan or its support, the sample is added slowly until the required weight is reached. The balance must be arrested before removing any of the substance from the container if too much is added in error.

9. Arrest the balance, after weighing, before removing the container and sample from the pan. Any spilled chemicals should be removed immediately from the pan or balance case with a small brush.

10. Close the case and return the weight knobs to zero.

Assignment

Practice weighing on the different types of balance available in the laboratory until you are confident you have mastered the correct technique.

9.5 Accuracy of balances

Most advanced analytical balances will weigh to considerably less than 1 mg. The accuracy of a 'fourth place balance', i.e. one which can be read to the nearest 10^{-4} g, is usually in the range of $\pm$ 0.05 to $\pm$ 0.1 mg.

Clearly the precision and accuracy of a top-pan balance for general purpose use in a laboratory, factory or workshop depends on its capacity. A typical figure for a balance weighing up to 200 g is about ± 0.01 g.

Assignment

Refer to the supplier's or manufacturer's leaflets and other literature and list the accuracy of the different types of balance in the laboratory.

9.6
Checking of balances

All balances should be checked regularly to ensure the continued accuracy of their readings. This check may be carried out using a box of standard weights (see sub-section 9.14 and fig. 9.15). These weights are manufactured to very small tolerances and should therefore be treated with extreme care. They should be transferred only with ivory-tipped forceps. They should never be touched with the fingers or with metal-tipped forceps, nor should they be allowed to come into contact with chemicals. The accuracy of a balance over its complete range may be checked by noting the balance readings as the standard weights are progressively added. All repairs and major adjustments to an accurate and sensitive analytical balance should be carried out by a specialist.

A convenient method for regular checking is to weigh an object on several balances and then to compare the readings. If the weights differ by a significant amount then the balances are in need of servicing.

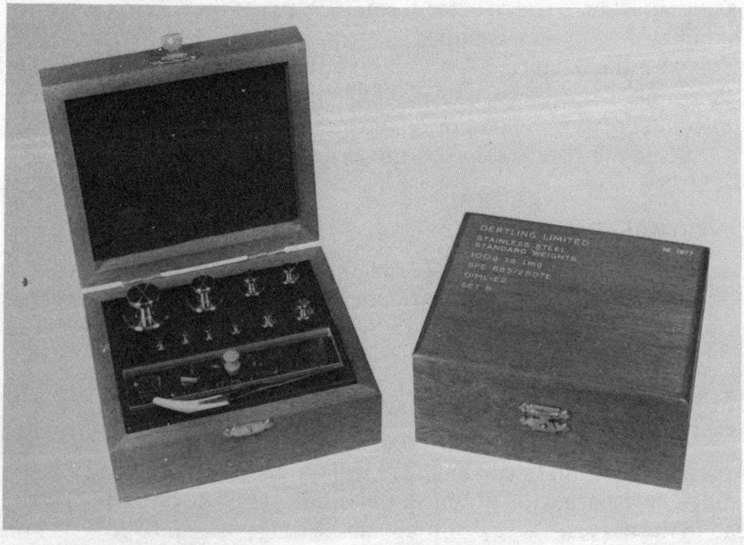

Fig.9.15 Standard weights

Assignment

Check two or more balances against one another by *carefully* weighing an empty stoppered weighing bottle or other convenient object on each. Write the weights down directly in your practical notebook (see 10.1). Avoid touching the weighing bottle or other object to be weighed with the fingers and remove the surface film of moisture or dust by polishing with a dry cloth.

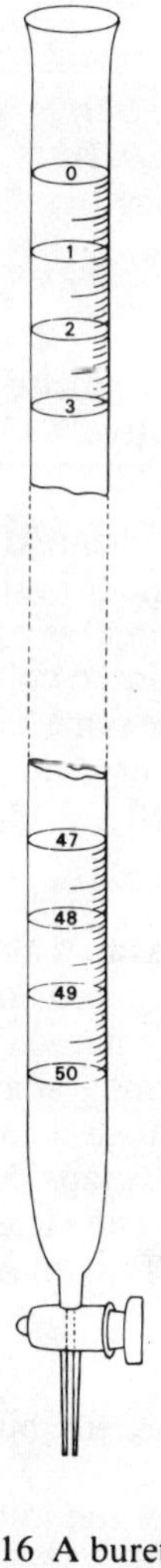

Fig.9.16 A burette

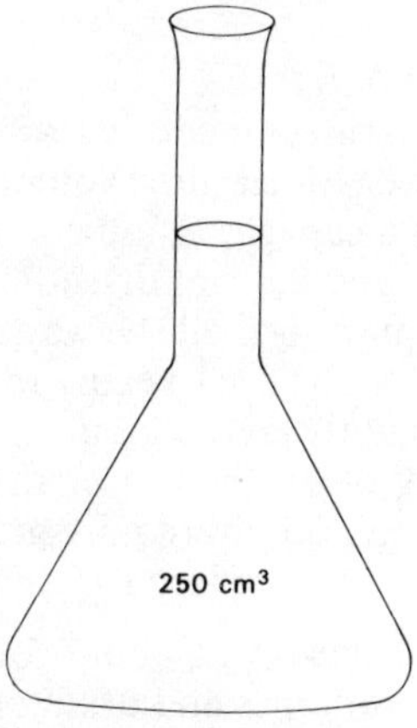

Fig.9.17 A volumetric flask

9.7
The use of a burette to deliver a stated volume of liquid

As its name implies, *volumetric glassware* is used for the precise measurement of volumes of a liquid. The *burette* (fig. 9.16) and the *pipette* (fig. 9.20) are designed to *deliver* known volumes of liquid, while the *standard* or *volumetric flask* (fig. 9.17) is made to *contain* a specified volume. The markings which frequently appear on volumetric glassware are listed in table 9.2.

Table 9.2 *Markings on volumetric glassware*

Marking	Meaning
A	Class A accuracy (see table 9.3)
B	Class B accuracy (see table 9.3)
Greenline	Coding for Class A accuracy
Goldline	Coding for Class B accuracy
In (or C)	contains
EX (or D)	delivers
15 sec (or 30 sec etc.)	drainage time for a delivery pipette
20 °C	temperature of calibration (see 9.13)

Volumetric galssware should always be thoroughly clean so that drops of liquid do not adhere to the inner walls. The methods available for cleaning such glassware (for example, with chromic acid) are described in sub-sections 11.1 and 11.2. Readings of all liquid volumes are taken at the bottom of the meniscus with the eye on the same level to avoid parallax errors (fig. 9.18(a)). Mercury surfaces curve upwards and levels are therefore read at the top of the meniscus (fig. 9.18(b)).

The burette
A burette is made by fitting a uniform graduated tube to a tap to deliver precisely measured volumes of liquid. The most commonly used burette has a capacity of 50 cm^3 (fig. 9.16). The graduations are marked in 1 cm^3 and 0.1 cm^3 divisions starting from the top. The 1 cm^3 gradations are inscribed all the way round the cylinder of the burette as an aid to adjust the eye level in order to avoid parallax errors. Glass taps should be lightly greased, but care should be taken not to use too much as this may block the tap or the capillary tip of the burette. Taps made of p.t.f.e. do not require greasing.

The procedure for using a burette is:

1. Rinse with about 5 cm^3 of distilled water, inverting the burette several times and drain out through the tap. Repeat.
2. Wash out the burette thoroughly with about 5 cm^3 of the solution and drain out through the tap. Repeat.

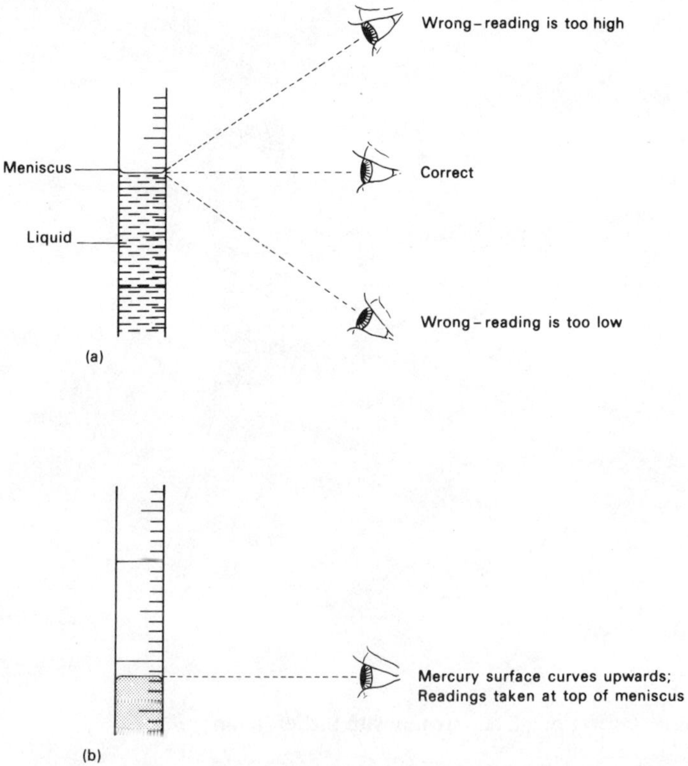

Fig.9.18 Reading liquid levels: avoidance of parallax errors

3. Fit the burette in the stand with burette clips and check that it is vertical.
4. Close the tap and fill the burette with the solution to 1–2 cm^3 above the zero mark using a funnel or a 100 cm^3 beaker and then remove the funnel from the top of the burette.
5. Open the tap to fill the capillary and drain the excess of liquid into the beaker to bring the bottom of the meniscus down to the zero.
6. Place the receiver (e.g. a conical flask) under the burette and, operating the burette tap with the thumb and index finger of the left hand as shown in fig. 9.19, dispense the required volume of the liquid. This method leaves the right hand free for mixing the solutions in the conical flask and also prevents leaks by pressing the tap into the barrel.
7. Allow the liquid to drain down from the inner walls of the burette before noting the final reading. The second decimal place may be estimated and recorded in brackets (e.g. 24.7(2) cm^3), but as this is much smaller than the allowed error (or *tolerance*) of the burette it is

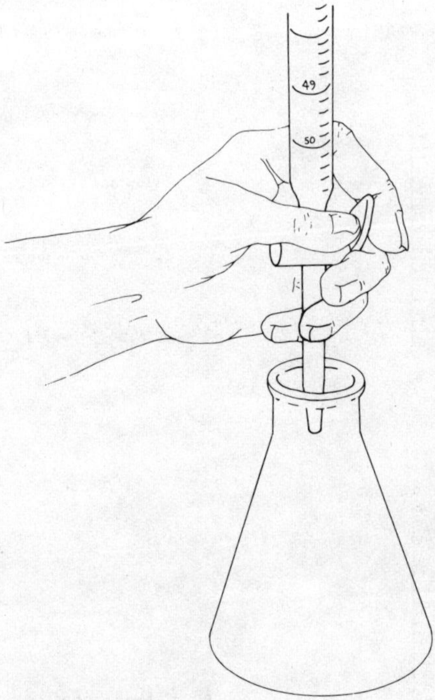

Fig.9.19 Operation of burette tap with the left hand

usually sufficient to record readings to the nearest 0.05 cm³. Care should be taken to collect the drop of liquid which may form on the end of the burette capillary in the receiver and to wash down the liquid clinging to the inner walls of the conical flask.

In a titration the procedure from step (4) would be repeated until consistent readings are obtained, e.g.

Titration	(1)	(2)	(3)
Final reading	22.40	44.75	22.35
Initial reading	0.00	22.40	0.00
Volume/cm³	22.40	22.35	22.35

The burette should be emptied and washed out thoroughly with distilled water after use. The procedure for cleaning volumetric glassware if further treatment is required is described in sub-sections 11.1 and 11.2.

9.8
Use of a pipette to deliver a stated volume of liquid

The main types of pipette in common laboratory use are shown in fig. 9.20. Figure 9.20(a) shows a transfer pipette which, as its name indicates, is used to measure out and transfer known volumes of liquid from one container to another. Figure 9.20(b) shows a graduated pipette used to measure and deliver small, precise volumes of liquid. The interval between successive graduations on a 5 cm^3 pipette, for example, is 0.05 cm^3 and liquid volumes are easily estimated to the nearest 0.01 cm^3.

The procedure for using a pipette is as follows:

1. Rinse out the pipette with a small amount of distilled water and allow to drain. Wipe the outside dry with a paper tissue or clean cloth.
2. Place the end of the pipette 3–4 cm below the surface of the liquid in the beaker and suck up about 5 cm^3 of the liquid. The use of a safety bulb or pipette filler (fig. 9.21) which eliminates the risk of swallowing dangerous liquids is recommended for this. Place the forefinger over the top end of the pipette and then, turning the pipette horizontally, rinse the liquid over the entire inner surface of the glass to about 1 cm above the graduation line. Drain the pipette and repeat.

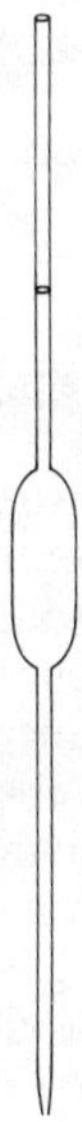

(a) A transfer pipette (a bulb pipette)

(b) A graduated pipette

Fig.9.20 Pipettes

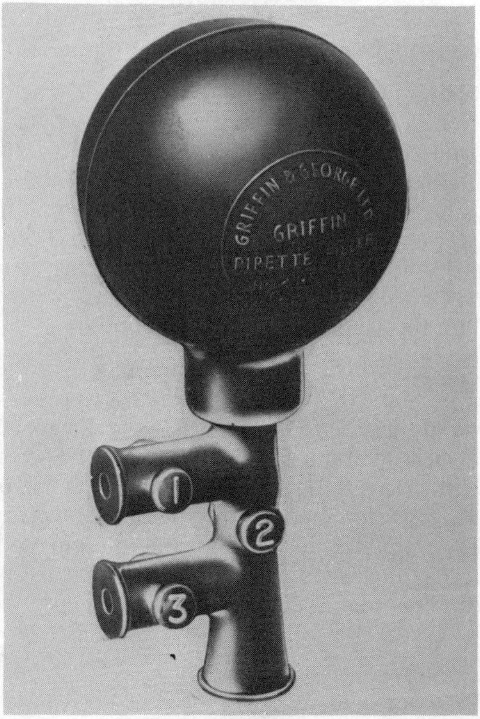

Fig.9.21 A pipette filler

3. Fill the pipette to a point just above the graduation line and press a dry forefinger over the top. Do not hold the pipette by the bulb in the centre as the heat of the hand will cause the liquid to expand.
4. Wipe the outside of the pipette with a tissue and then, with the eye level with the graduation line, slowly release the pressure of the forefinger and allow the liquid to drain back into the beaker drop by drop until the bottom of the meniscus just touches the graduation mark. Touch the tip of the pipette against the inner wall of the beaker to remove any solution adhering to the end of the pipette.
5. Allow the pipette to drain into the receiving vessel (e.g. a conical flask) and then, after a few seconds (3 s for those marked IN or C, 15 s for those marked EX or D—in some cases, e.g. with graduated pipettes, the drainage time is marked on the pipette) touch the tip of the pipette against the inner wall of the flask. *Do not* blow out the liquid remaining in the pipette; allowance for it has been made during the calibration of the pipette.

Note: Although it is safe (but not hygienic) to pipette many solutions by mouth, this practice should be discouraged. Pipette fillers or safety bulbs MUST be used for pipetting concentrated acids, strong alkalis, corrosive liquids, poisons, solutions of radioactive materials, bacterial suspensions and biological samples or volatile solvents.

Pipettes should be washed out thoroughly with distilled water after use. The procedure for cleaning them is described in sub-sections 11.1 and 11.2.

9.9
Accuracy of volumetric glassware

Volumetric glassware is manufactured in three different standards of accuracy. The maximum permitted error (or tolerance) for the two graded classes A and B are fixed by the British Standards Institution (see 9.16) while the third class is ungraded. Grade A glassware is available with a works examination certificate stating the precise volume. The BSI tolerances for Class A and Class B burettes, pipettes and volumetric flasks are given in table 9.3. Class A (Greenline) glassware is two or three times more expensive than Class B (Goldline) equipment. The accuracy of Class B glassware is better than 0.4%, which is well within the limits of experimental accuracy expected in schools and so the less expensive Class B glassware is generally suitable for educational use. However, for more advanced classes and for critical analytical work Class A (or Greenline) apparatus is essential.

Table 9.3 *BSI tolerances for volumetric glassware*
(a) *Burettes* (*BS 846 : 1962*)

Nominal capacity /cm³	Subdivision /cm³	Tolerance /cm³ Class A	Class B
25	0.1	± 0.05	± 0.1
50	0.1	± 0.05	± 0.1

(b) *Pipettes* (*BS 1583 : 1961*)

	Nominal capacity /cm³ 5	10	20	25	50
Tolerance (± cm³) Class A	0.015	0.02	0.03	0.03	0.04
Tolerance (± cm³) Class B	0.03	0.04	0.06	0.06	0.08

(c) *Volumetric flasks* (*BS 1792 : 1952*)

Nominal Capacity	Tolerance	
	Class A	Class B
	± /cm³	
50	0.04	0.06
100	0.06	0.1
250	0.1	0.2
500	0.15	0.3
1000	0.2	0.4

9.10
The multimeter

A multimeter is a multipurpose instrument for measuring electrical quantities. By turning a switch it is possible to set the meter to measure alternating current (a.c.) or direct current (d.c.). It may then be converted into an ammeter for measuring current (amperes), a volt-

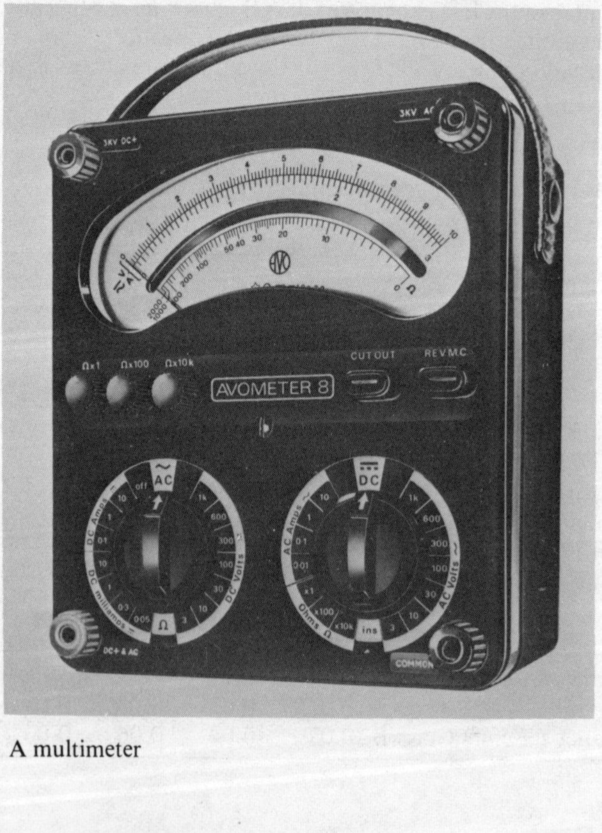

Fig.9.22 A multimeter

Fig.9.23 A digital multimeter

meter for measuring potential difference (volts) or an ohmmeter for measuring the resistance (ohms) of a component or circuit. A multimeter may be set to different ranges, e.g. 0–100 mV and volts × 3, 10, 30, 100, 300 and 1000. This enormously extends its use and makes it a versatile, portable instrument for electrical installation work, fault finding and general laboratory use in electronics and electricity. A multimeter is fitted with an automatic cutout device and fuse to protect it from damage caused by accidental overloading. Nevertheless, if the approximate magnitude of the current or voltage to be measured is unknown, the meter should be set first to one of the higher ranges. The range switch is then turned back until a third to full scale deflection reading is obtained. Examples of analogue and digital multimeters are shown in figs. 9.22 and 9.23. The analogue model has a mirror mounted on the scale. Parallax errors are eliminated by taking readings at the point where the needle is directly superimposed on its mirror image. Most multimeters are fitted with a Weston standard cell (see 9.13) which provides an accurate and extremely stable reference for its measurements. The meter is essentially a moving-coil microammeter in which resistors may be connected in series or in parallel to convert it into a voltmeter or into an ammeter. A small battery is fitted as a power source for measurements of resistance and of continuity within a circuit. The instrument is first set to zero after connecting its terminals together before measuring resistance. Care must be taken when reading the resistance scale on the analogue version of the multimeter (fig. 9.22), as (unlike the current and voltage scales) this scale is not linear. An example of the use of a multimeter to check continuity is the testing of cartridge fuses (see 1.5). A very high ('infinite') resistance reading indicates that the circuit is not continuous and that the fuse has blown.

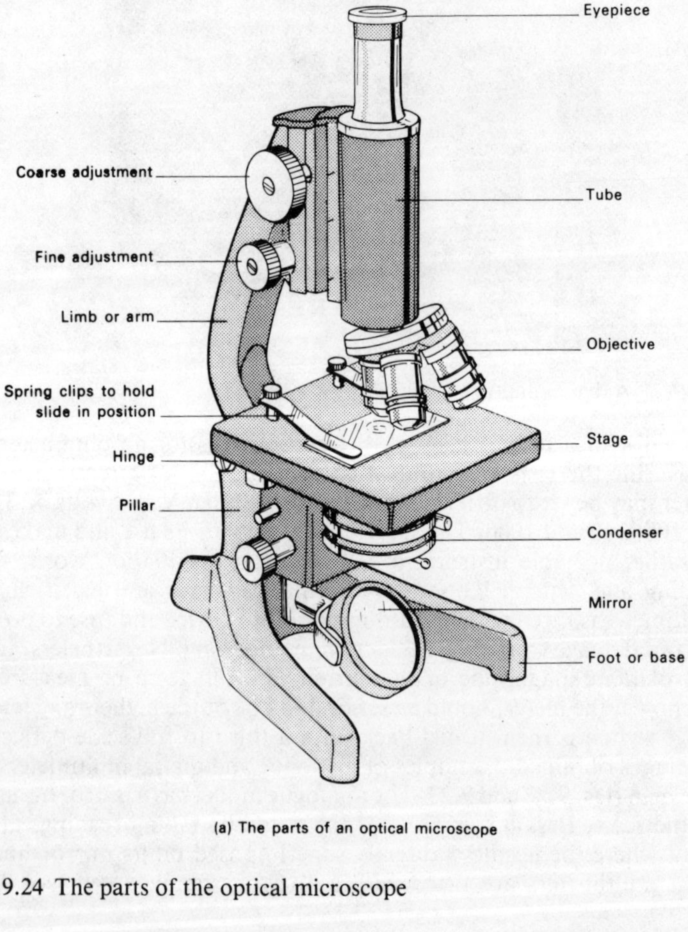

(a) The parts of an optical microscope

Fig.9.24 The parts of the optical microscope

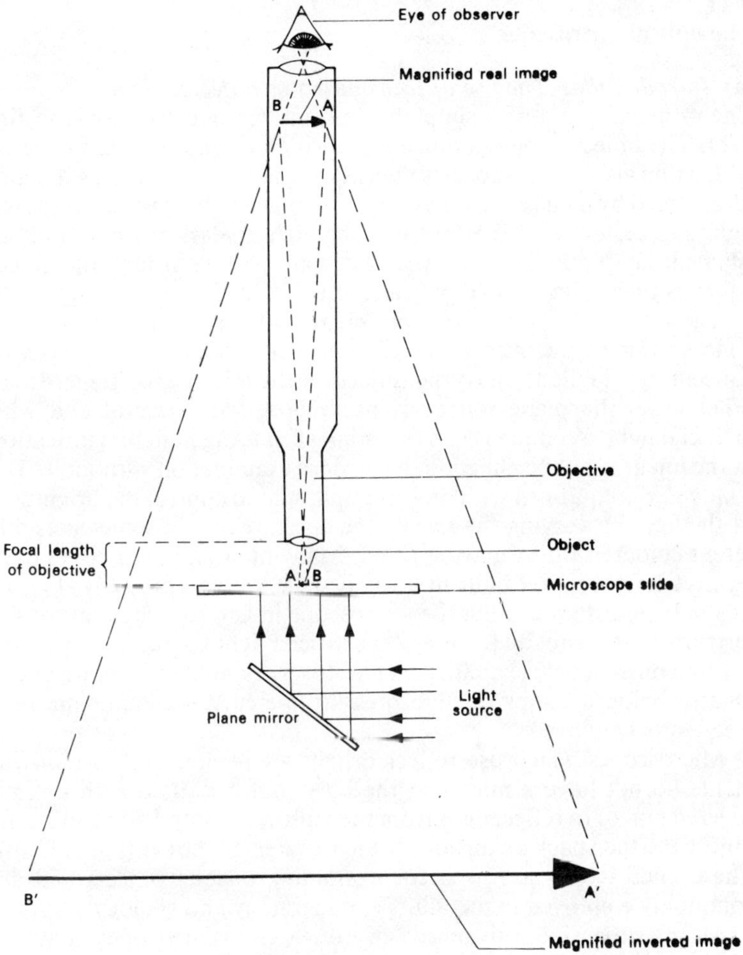

Fig.9.24 (b) Ray diagram of the optical microscope

Assignment

Use a multimeter to measure a.c. and d.c. potential and to test for continuity. Measure the resistance of a number of colour coded resistors and compare the readings obtained with the coded values and tolerances listed in table 1.2.

9.11
The optical microscope

(a) *Introduction: The use of transmitted or reflected light*
The principal parts of a simple optical microscope are shown in fig. 9.24. The object to be examined is placed on a glass slide and covered with a thin glass cover slip. It is then placed on the microscope stage and illuminated by light reflected by the mirror. Daylight may be used as the light source, but if this is not light enough a pearl or opal tungsten filament lamp may be employed. The light passes through the object which is thus being viewed by transmitted light.

The mirror has two sides: one plane and the other concave. The concave side of the mirror should be used if there is no condenser to concentrate the light on to the object on the microscope stage. In all other cases the plane mirror is used, both with natural and with artificial light. A condenser is essential when using high magnification as the image is not bright enough for close examination without it. The condenser is equipped with an iris diaphragm to control the brightness of the light by varying the size of the aperture and in some cases different coloured filters may be fitted. Many modern microscopes have an electric light source built-in. Provided the source is properly aligned, it is only necessary to plug the microscope in and to switch on for the instrument to be ready for use. An artificial light source must be used for fine microscopic examination and for work at high magnification. Quartz–halogen lamps or high pressure mercury arc lamps are now frequently employed.

Microscopes which use reflected light are available. These instruments do not have a mirror at the base, but are fitted with a semi-silvered mirror to reflect light from the bulb in the barrel down on to the object and then back up through the lens system to the eye (fig. 9.25(b)). These microscopes are used for examining opaque objects and are commonly employed in metallurgy, mineralogy and geology.

More expensive advanced or research microscopes have a mechanical stage fitted with vernier scales to indicate a position on the slide so that the required area may be easily located and examined at a later date (fig. 9.26). A binocular microscope in which the light rays coming from the objective are split by a prism so that half the light is transmitted to each eye is shown in fig. 9.27. Many specialised techniques are available for improving contrast and visibility of the magnified image or for increasing the amount of information which may be obtained from the microscopic examination of the material. These techniques include dark-ground microscopy, selective staining, the use of oil immersion objectives to obtain very high magnification, phase contrast microscopy and the polarizing microscope.

To improve the clarity of the image obtained with objectives of the highest magnification (i.e. those with focal lengths less than about 2

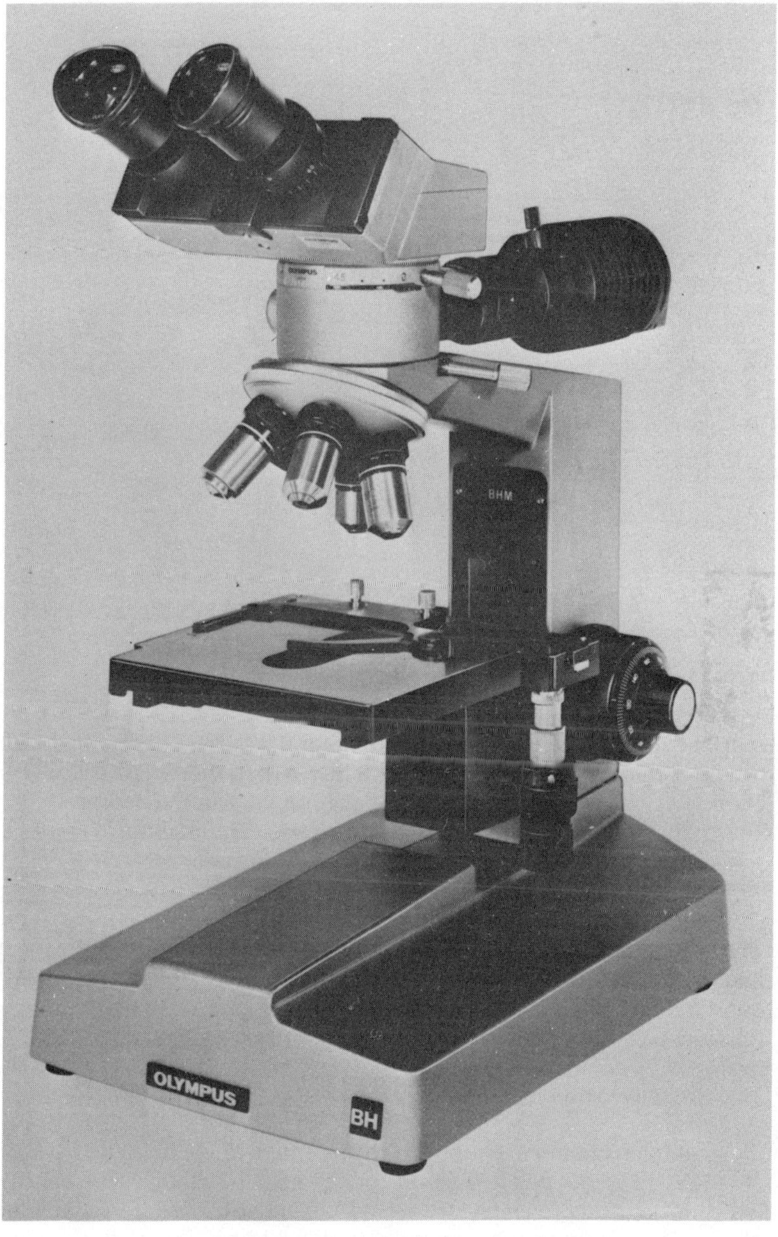

Fig.9.25 (a) A reflecting (metallurgical) microscope

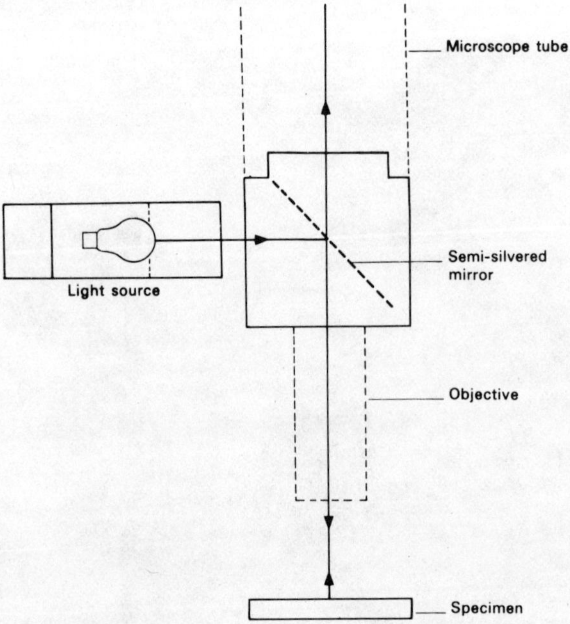

Fig.9.25 (b) The reflecting microscope

mm) the space between the objective lens and the cover slip is filled with a drop of oil with a similar refractive index to that of the glass slide. This eliminates light loss from refraction or total internal reflection. Cedar wood oil is commonly used for this purpose. The highest magnifications in table 9.4 are obtained using an oil immersion objective.

(b) *Magnification*
The magnification of a particular microscope is determined by the focal length of the objective and the magnification of the eyepiece. An approximate value of the magnification may be calculated from the relationship:

$$\text{magnification} = \frac{\text{tube length} \times \text{magnification of eyepiece}}{\text{focal length of objective}}$$

Most microscopes are supplied with a range of eyepieces, e.g. × 5, × 8, × 10, and × 12. The objectives may be changed by simply twisting the nosepiece (fig. 9.24) and clicking another objective into place. The approximate magnifications obtained by various combinations of a number of common objectives and eyepieces are listed in table 9.4. The microscope is assumed to have the standard tube length of 160 mm.

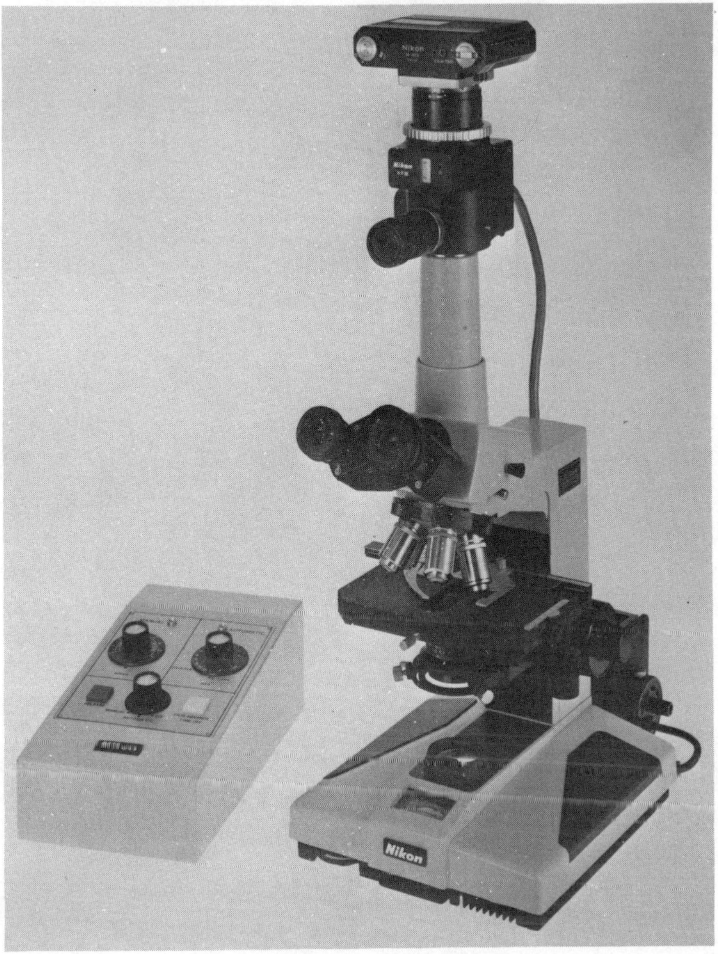

Fig.9.26 A sophisticated research microscope equipped with a vernier mechanical stage

(c) *Procedure for using a light microscope*
The steps for setting up and using a simple microscope with an external light source are:

1. Holding the microscope by its limb, remove the instrument from its case and place it in front of a window or 15–20 cm from the opal bulb of an electric lamp.
2. Position the slide on the stage so that the object is above the centre of the condenser.

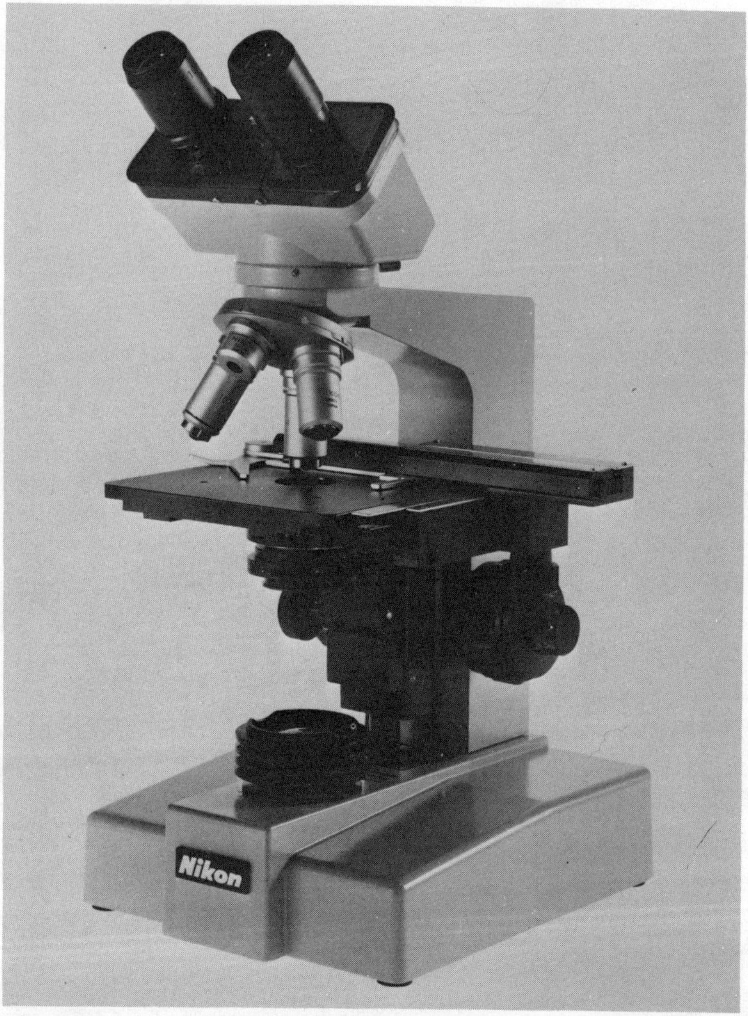

Fig.9.27 The binocular microscope

3. Click the lower power objective (16 mm or $\frac{2}{3}$ in) into position.
4. Adjust the mirror so that the object is evenly illuminated.
5. Using the coarse and fine adjustment knobs, lower the objective until its lens is just above the cover slip on the microscope slide. Slowly raise the objective by moving the coarse adjustment knob in the opposite direction until the object is roughly in focus. If the light is too bright, adjust or close the substage iris diaphragm to reduce glare. The light intensity may also be reduced by inserting suitable filters in the light path or by regulating the electrical supply to the bulb with a rheostat or variable resistance.

6. Use the fine adjustment knob to bring the image sharply into focus. It is best to keep both eyes open when using a microscope. This is less tiring than working with one eye closed, although it does require a little practice to reject the images coming to the free eye. People who wear glasses should remove them before using a microscope.
7. Remove the slide and return the microscope to its case after use.

Table 9.4 The optical microscope: Magnification of some lens systems
Total magnification = primary magnification × magnification of eyepiece. Primary (1°) magnification = tube length *l*/focal length *f* of objective lens. Standard microscope tube length *l* = 160 mm

Focal length of objective *f* /mm	/in	1° magnification *l*/*f*	Magnification obtained Eyepiece No. 1 × 5	No. 2 × 6	No. 3 × 8	No. 4 × 10	No. 5 × 12	No. 6 × 15
25	1	× 6.4	32	38	51	64	76	96
16	$\frac{2}{3}$	× 10	50	60	80	100	120	150
8	$\frac{1}{3}$	× 20	100	120	160	200	240	300
6	$\frac{1}{4}$	× 26.6	133	160	213	266	319	399
4	$\frac{1}{6}$	× 40	200	240	320	400	480	600
3	$\frac{1}{8}$	× 53.5	265	318	424	530	636	795
1.5 (oil immersion)	$\frac{1}{16}$	× 115	575	690	920	1150	1380	1725

Assignment

Examine a simple optical microscope employing either reflected or transmitted light. Compare it with the diagram in fig. 9.24 and make certain you know the name and function of all its principal parts. Use the microscope to examine a number of prepared slides until you are proficient in its use.

9.12
Correct procedure for cleaning a microscope

A microscope should always be covered, preferably with a polythene sheet or bag, or returned to its case when it is not in use. Under no circumstances should a microscope be left with the eyepiece removed as this allows dust to settle inside the instrument and on the inner surfaces of the objective lens.

A microscope may be cleaned using the following procedure:

1. Remove any dust or dirt from the microscope stage with a brush or with a paper tissue moistened with xylene. (*Note*: Never use alcohol for this purpose or for cleaning lenses.)

2. Dust the mirror and the condenser and the eyepiece and objective lenses with a camel hair brush or with a lens tissue. A jet of air from a blower as used for cleaning camera lenses is often useful for dislodging dust particles from inaccessible places. The location of a speck of dust in the optical system may be found by rotating the eyepiece while looking through the instrument. If the dust is on the eyepiece the speck will move too, but if the speck remains stationary the dust is probably on one of the surfaces of the objective lens or perhaps on the lens of the condenser.
3. Lightly grease the coarse and fine adjustments and the slides of the mechanical stage according to the maker's instructions.
4. Finally, polish the mirror and lenses with a clean lens tissue. Do not use paper towels or cleaning rags for this purpose as these will damage the lens surface. Remember, a microscope is a delicate, precision instrument and must be treated with care. The repair of any optical faults or mechanical troubles should be left to a skilled microscope mechanic.

9.13
Primary standards for physical, chemical and biological quantities

The SI units of such physical quantities as length, time, mass, thermodynamic temperature and current are defined in terms of invariable, accurately reproducible quantities. For example, the metre is defined as the length which is equal to 1 650 763.73 wavelengths in a vacuum which correspond to a particular transition in the orange radiation from gaseous krypton-86 atoms. Similarly, the unit of time, the second, is defined in terms of an exact number (9 192 631 770) of periods of the radiation corresponding to a particular transition in the caesium-133 atom. The caesium clock is accurate to less than one second in 3000 years. The definitions of other SI units are listed in Appendix I.

All standards are related back to precisely defined quantities such as these. These standards are known as *primary standards*. The primary standard of mass, for example, is the cylinder of platinum–iridium alloy stored in Sevres, near Paris. By definition, its mass is precisely 1 kilogram.

9.14
Secondary standards

In many cases primary standards do not provide a convenient, *direct* method of standardisation in the laboratory and practical *secondary standards* and duplicates based on the primary standards are therefore employed. These standards are manufactured to the required degree of accuracy and—provided they are treated carefully or are made up to the exact specification—they will give a reliable method of calibrating instruments or of measuring physical, chemical and biological quantities. Many of these measuring instruments or standards are available from the suppliers with a National Physical Laboratory (NPL) or British Standards Institution certificate specifying their accuracy. Examples include thermometers, standard masses, standard Weston cells and gauge blocks. A technician should be familiar with the practical standards employed for the quantities commonly used in his own laboratory. For example:

Length—accurately machined gauge blocks of known dimensions may be used to standardise micrometers and other measuring instruments (see table 9.5).

Table 9.5 Tolerance of Grade O metric gauge blocks

Range /mm	Tolerance/measuring accuracy
20–60	0.000 15
60–80	0.000 20
80–100	0.000 25

Mass—a box of standard weights may be used to check the accuracy of a balance (see fig. 9.15 and sub-section 9.6)

Time—the ready availability of quartz clocks and watches accurate to within a few seconds per month provides a convenient standard for correcting clocks and stopwatches. The quartz clock may be set and checked using the Greenwich time signal.

Voltage—The *Weston standard cadmium cell* is the universally adopted standard for precision electrical measurements. It has long term stability and varies very little with temperature. The cell usually consists of an H-shaped vessel (fig. 9.28) which contains the chemicals of the highest available purity. The electromotive force (e.m.f.) of the cell is 1.018 59 volts at 20°C and the temperature coefficient is $-40\ \mu V$ per °C. The cell is often mounted in a metal case filled with oil to ensure that the two arms of the cell are at the same temperature. Weston standard cells are damaged if a current is withdrawn from them and it is therefore advisable to place a very high resistance, of about 10 000 Ω, in the circuit to guard against accidental current drain.

Disintegration rate—standard radioactive sources of known disintegration rate or specified source strength of gamma-radiation, for example, are available from the Radiochemical Centre, Amersham, for calibrating the ratemeters of Geiger counters etc.

pH—The primary standard is a 0.05 molar aqueous solution of pure potassium hydrogen phthalate (see BS 1647 : 1961). By definition, this solution has a pH of 4.000 at 15 °C. Its pH at any temperature between

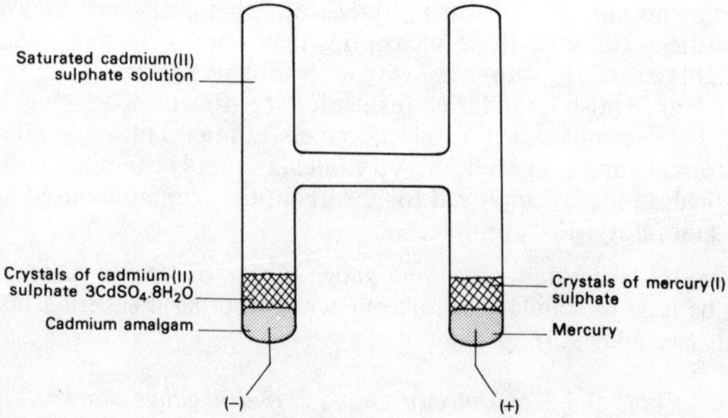

Fig.9.28 The standard Weston cell

O and 55 °C is defined by the formula:

$$pH = 4.000 + \frac{1}{2}\frac{(t - 15)^2}{100}$$

This buffer solution provides a convenient standard for the calibration of pH -meters for use in both chemical and biological laboratories.

Other commonly used practical standards for physical, chemical and biological quantities are summarised in table 9.6.

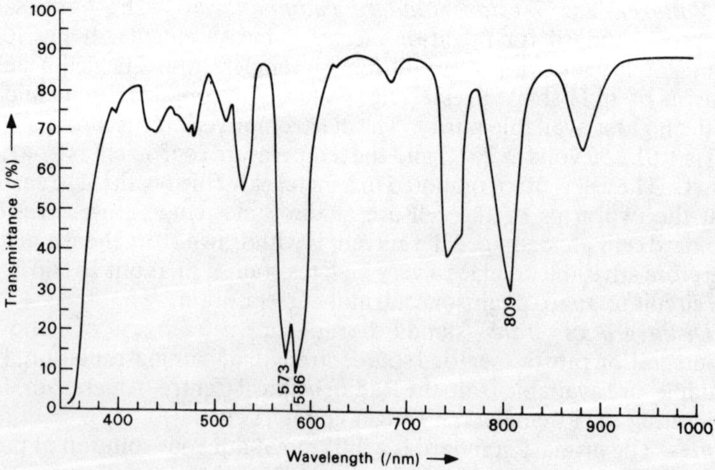

Fig.9.29 Absorption spectrum of didymium

Table 9.6 Common practical standards for physical, chemical and biological quantities

Quantity	Purpose of Standard	Practical standard employed
Conductivity	Calibration of conductivity cells	Aqueous potassium chloride solutions of known concentration (see Appendix VII).
Light intensity	Calibration of photocells	Gas filled, tungsten filament standard bulbs.
Concentration	Standardisation of solutions in volumetric analysis	Primary analytical standards (e.g. sodium carbonate, borax, benzoic acid, sodium chloride, sodium oxalate, potassium iodate), i.e. stable substances which are readily available in the pure state and which can be used to prepare solutions of known concentration by direct weighing. The principal primary standards for volumetric analysis are listed in Appendix VIII.
Wavelength	Calibration of visible range spectrophotometers	A didymium filter. The sharp peaks at 573, 586 and 809 nm in the absorption spectrum of didymium (see fig. 9.29) are used to calibrate the wavelength scale.
Optical density	Calibration of photometric scales of u.v. and visible spectrophotometers	Aqueous potassium chromate solution of known concentration (see Appendix IX).
Wavelength or wave number	Calibration of infra-red (i.r.) spectrophotometer	The known wavelengths (or wave numbers, where wave number $\nu(/\text{cm}^{-1}) = \frac{1}{\text{wavelength } \lambda\ (/\text{cm})}$

Table 9.6 *(Continued)*

Quantity	Purpose of Standard	Practical standard employed
		of sharp peaks in the i.r. absorption spectrum of polystyrene (see fig. 9.30) are used.
Volume	Calibration of burettes, pipettes and standard (volumetric) flasks	Weighing measured volumes of distilled water. The volume delivered by the pipette or burette or contained by the flask is calculated from the density of distilled water at the temperature of the determination (see Appendix X). Examples of such calibrations for a burette and pipette are shown in fig. 9.31 and sub-section 10.1(c) respectively.

Assignment

Make a list of the quantities and their standards which apply to the instruments you commonly use at work. Reference to suitable specialised texts and to publications such as the *British Standards Yearbook* (see 9.16 and 9.17) may be necessary to carry out this assignment.

9.15
Periodic testing of common laboratory instruments against standards

It is not possible to give a general schedule for the testing of all the different types of laboratory equipment against the standards described in sub-section 9.14. In some cases this schedule in determined by the instrument, for example a pH meter has to be set with a buffer of known pH (e.g. a 0.05 mol/l solution of potassium hydrogen phthalate, pH = 4.000 at 15 °C—see 9.14) immediately before it is used, while other equipment, such as a conductivity cell, need be calibrated only if the original calibration figure has been lost or the cell damaged or if a series of extremely accurate measurements are to be taken as part of an

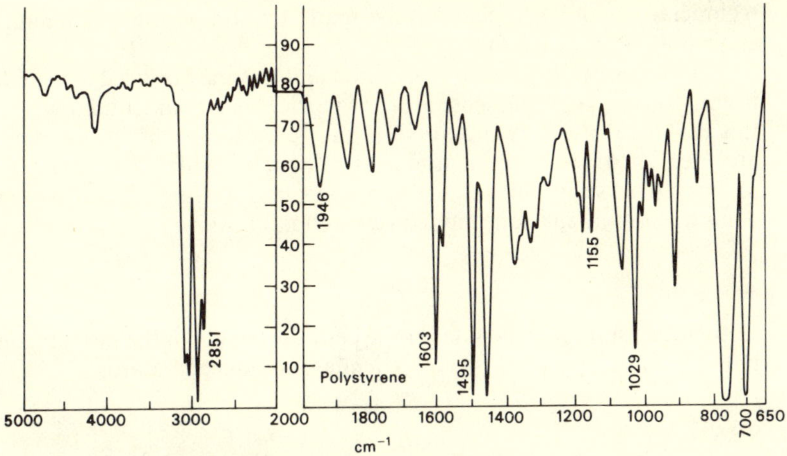

Fig.9.30 Infra-red absorption spectrum of polystyrene

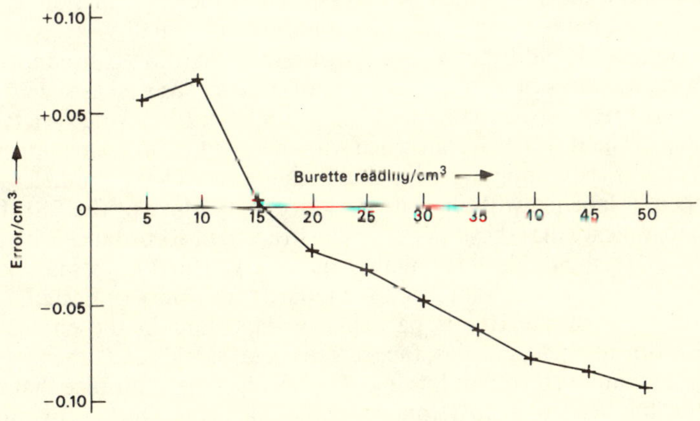

Fig.9.31 Calibration graph of a 50 cm^3 class A burette

experiment. Similarly, the stability of a multimeter (see 9.10) is so high that calibration is rarely required, even over a period of many years. Testing is required only for highly critical work and even then an annual confirmation of accuracy is sufficient.

In other cases, such as the checking of analytical balances, the frequency of testing is determined partly by the accuracy to which readings are required and partly by the amount of use (or misuse) the instrument receives. A few weighings by a particularly clumsy or messy operator can produce more harm than a hundred weighings by a skilled technician. A quick check may be made by comparing one balance against another (see 9.6).

In most cases, the maker's recommendations may be used as a guide in constructing a testing schedule for a particular instrument. However, this may be drastically modified by the age, use, permitted tolerance and location of the equipment. An instrument in regular use in a factory or busy workshop will require more frequent testing than the same equipment in an air-conditioned research laboratory.

Assignment

Construct a suitable schedule for the periodic testing of the instruments in the laboratory where you work against suitable standards.

9.16
Purpose of the BSI

The British Standards Institution was founded in 1901 and has operated under a Royal Charter since 1929 as the sole national standards organisation of the United Kingdom. Its principal function is to prepare and make known national standards for quality, safety and performance. It publishes glossaries of terms, definitions, quantities, symbols, methods of testing and codes of practice, as well as specifications for safety, performance and dimensions etc. for a wide variety of products. The BSI is also concerned with the certification and approval of products as complying with these standards. These products are stamped or labelled with the Kitemark or Safety Mark of the BSI (fig. 9.32) to indicate that they have been constructed in accordance with the relevant British Standard for quality and safety. The BSI does not have powers to enforce the adoption of standards for safety or design.

The work of the BSI is particularly important in the choice of respirators, helmets, goggles, face screens, heat shields and other safety equipment and protective clothing. The user has the assurance that the product has been made to a standard which will protect him from injury in the event of an accident. Price alone is not always a dependable

The kitemark

The safety mark

Fig.9.32 Kitemark and Safety Mark of the British Standards Institution

indication of quality and performance, nor for that matter is the salesman's recommendation. An accident is not the time to discover design or performance defects in equipment intended for protection.

The various British Standards for defining and determining the accuracy of instruments or procedures for carrying out specified operations are also of interest to technicians. The titles of a number of these are given in sub-section 9.17.

All British Standards are identified by the initial letters BS and a serial number which is usually followed by the year of publication. For example, BS 349 : 1973 refers to the colour coding for the identification of the contents of industrial gas cylinders. Other BSI publications are identified by the appropriate initial letters, e.g. CP and DD for 'Code of Practice' and 'Draft for Development' respectively. Examples of these are given in sub-section 9.17. A number of European standards, denoted by the initial letters EN, have now been introduced. EN 2 : 1972 (BS 4547), for example, defines the code for the classification of fires (see 5.1(b)).

A full list of British Standards and other publications in current use is given in the *British Standards Yearbook*, which should be available for reference in virtually all college libraries as well as many public or industrial libraries. The equivalent organisation in the United States is the American Standards Association. Its standards are identified by the letters ASA.

Assignment

Examine a number of items of equipment at home and at work and list those which carry the Kitemark or Safety Mark of the BSI.

9.17
Titles of useful BSI publications

Every technician should know the titles of the BSI publications which refer to his or her own field of work. It is not possible to list all of these as the range of requirements of technicians working in food science or medical laboratories, for example, to those working on chemical plant, in research establishments or in school laboratories is enormous. The reader is therefore recommended to consult the *British Standards Yearbook* for a list of all the standards and publications currently in use and to note down the titles and consult those which apply to his own work.

The following standards have been referred to in this book or are of general interest:

BS 700 : 1976	Specification for graduated pipettes.
BS 846 : 1962	Tolerances of burettes of capacities 1–100 cm^3.
BS 870 : 1950	External Micrometers.

BS 887 : 1950	Vernier callipers.
BS 889 : 1965	Flameproof electric light fittings.
CP 1003 : 1964, 1966	Electrical safety in explosive atmospheres.
CP 1013 : 1965	Methods of earthing electrical equipment.
BS 1361 : 1971 and 1362 : 1973	Cartridge fuses for plugs.
BS 1583 : 1961	Specification for one-mark pipettes, including tolerances for Class A and Class B glassware.
BS 1647 : 1961	The pH-scale.
BS 2091 : 1969	Respirators for protection against harmful dust, gases and agricultural chemicals.
BS 2092 : 1967	Industrial eye protectors.
BS 2586 : 1965	Glass electrodes for pH measurement.
BS 2769 : 1964	Portable electrical tools.
CP 3013 : 1974	Fire precautions on chemical plant.
BS 3145 : 1978	Specification for laboratory pH-meters (includes maximum permitted overall error and errors arising from twelve sources plus test methods for assessing electrical performance).
BS 3664 : 1963	Film badges for monitoring radiation.
BS 3996 : 1978	Colour coding for pipettes.
BS 4311 : 1968	Specification for metric gauge blocks.
BS 4402 : 1969	Safety of laboratory centrifuges.
BS 4667 : 1974	Escape breathing apparatus.
BS 5142 : 1974	Standard cells
BS 5378 : 1976	Specification for safety colours and safety signs.
BS 5423 : 1977	Specifications for portable fire extinguishers.
BS 5686 : 1976	The use of SI units (see also BS 3763 : 1976).
DD 48 : 1976	Identification of fire extinguishers.
DD 52 : 1977	Recommendations for the presentation of tables, graphs and charts.

Assignment

Compile a list of the BS publications which apply to your own field of work.

Questions: 9 *Instruments*

9.1 Describe how the vernier and micrometer scales 'magnify' small size differences and thus make it possible to measure distances to a far greater precision.
9.2 What is the difference between precision and accuracy? Is it possible for a highly precise measurement to be completely inaccurate?
9.3 Why is it important to treat vernier slide callipers and micrometer screw gauges with care and not to drop them on the floor, for example?

9.4 Sketch a micrometer screw gauge and name its important parts.
9.5 Why is a micrometer gauge usually fitted with a spring ratchet mechanism?
9.6 Why is it important to check the zero reading of a micrometer screw gauge before taking measurements? How is the zero reading used?
9.7 State the expected accuracy of measurements made with a vernier and a micrometer.
9.8 Describe how you would weigh out accurately a sample of a powdered chemical. Why can't this sample be weighed out on a filter paper or directly on the scale pan?
9.9 State the expected accuracy to which the mass of a sample of about 1 g may be determined using an analytical balance.
9.10 Why is it important to clean up any spilled chemicals immediately from the scale pans or balance case when weighing?
9.11 What information is obtained by comparing the mass of an object weighed on two or three different balances?
9.12 Describe the procedure for using (a) a burette and (b) a pipette to deliver a stated volume of liquid.
9.13 Why is it important that volumetric glassware should always be clean before use?
9.14 Why is it important to remove the funnel from the top of a burette after filling it?
9.15 Is it correct to blow out the last drop of liquid from a pipette?
9.16 Describe how you would check the calibration of a burette by weighing measured volumes of pure water.
9.17 Why is it important for the eye to be on the same level as the bottom of the meniscus when reading the volume of a liquid?
9.18 Why is it advisable not to pipette liquids by mouth?
9.19 What is the accuracy of the volumes of liquid (say 25.00 cm^3 samples) delivered by a burette or pipette?
9.20 What is a multimeter? How would you use a multimeter (a) to test for continuity in a circuit and (b) to measure a.c. and d.c. potential?
9.21 Why is it best to use an electric lamp as the light source for an optical microscope rather than daylight for work at high magnification?
9.22 What is an optical microscope employing reflected light used for?
9.23 Why is it important not to leave a microscope with the eyepiece removed for any length of time?
9.24 What is the difference between a primary and secondary standard?
9.25 Why is it necessary to check instruments against suitable standards from time to time?
9.26 Name *three* practical standards for common chemical, physical or biological quantities.
9.27 Name *three* factors which determine the frequency with which measuring instruments may require calibration.

9.28 What are the functions of the British Standards Institution? Draw diagrams of the BSI Kitemark and Safety Symbol.
9.29 Why is it important for the BSI to be independent of industry for its financial support?
9.30 State the titles of *three* BSI publications which apply to your own field of work.
9.31 Do you think that the BSI should have power to enforce the adoption of standards?

D Scientific reporting

Section 10: *The expected learning outcome of this section is that the student should be able to understand the necessity for clear, accurate, honest reporting of experimental results*

Specific objectives: *The expected learning outcome is that the student:*
10.1 Records all experimental observations in a permanent notebook.
10.2 Reports experimental results, despite preconceived ideas.
10.3 Chooses the most appropriate form in which to present experimental results: tabular, graphical, pictorial.

10.1
The practical notebook

All scientists and technicians need to be able to communicate effectively both verbally and in writing for a variety of purposes. It may be necessary to report the results of a series of investigations which have been carried out or to contact a firm either directly or through your immediate superior or section leader about the replacement or repair of an instrument. But communication—however important it may be—is not the sole purpose of scientific writing. We also write things down to help us remember them (as in making notes during a lecture) and to help us to observe and to plan and organise material.

Scientific reporting is especially important in laboratory work. All observations made during an investigation should be recorded *directly* and *immediately* in a bound laboratory or field notebook. It's no use writing the observations on scraps of paper or on the back of a cigarette packet where they can easily be lost, nor should the technician rely solely on his memory. The record of all the experiments you carried out as part of the coursework of a TEC unit at college is also valuable for revision purposes.

The following procedure is recommended for recording experimental observations in a practical notebook:

1. Write your name, the subject and course number in capital letters on the cover and your name and home address on the front page. It is also worthwhile to add the name and section of the firm where you work as well as the name and department of the college you attend.
2. Number each page in the notebook and reserve the first two or three pages for a list of contents.

3. Always write in ink or with a ballpoint pen so that all entries are permanently legible.
4. All entries should be made directly into the notebook and not on scraps of paper with the intention of copying them later.
5. Start each experiment on a new page and always begin with the date and title of the experiment. Enter the page number and title of the experiment in the contents list at the beginning of the book.
6. Write up the various sections of the experiment under the appropriate headings: Introduction and/or Aim of the Experiment. Apparatus and Materials. Method. Readings/Observations/Measurements. Calculation. Results. Discussion. Conclusion. References.

The apparatus, materials and method sections should be sufficient to enable a student on the same course to repeat the experiment successfully and obtain satisfactory results. A labelled diagram of the apparatus should be included if it is appropriate and, wherever possible, data should be organised into tables or on carefully prepared data sheets which are then securely attached to the relevant page of the practical notebook. Each column in the data table should be headed to show what the figures refer to and the units of measurement should be given (see 10.3). All significant raw data should be included: for example, when weighing 1.2—1.3 g of sodium carbonate to prepare a standard solution for titration against an acid, the mass of the empty weighing bottle (m_1) and the mass of the weighing bottle plus the sodium carbonate (m_2) should both be recorded as well as the accurate mass of the sodium carbonate ($m_2 - m_1$).

An example of the recommended method of writing up a laboratory report is given in a later part of this section.

(a) *Significant figures*

An indication of the precision of a measurement should be conveyed by the number of significant figures in its numerical value. This figure should include all the certain digits plus the final estimated or uncertain digit. A measured value of 25.1, for example, indicates that the value is closer to 25.1 than it is to either 25.0 or 25.2, and that the uncertainty in the measurement is ± 0.1 unit. The relative uncertainty is thus 1 part in 251 or approximately 0.4%. The relative uncertainty if the value were expressed as 25.10 ± 0.02 would be 2 parts in 2510 or about 0.08%.

All the certain figures plus the first uncertain figure should be recorded when reporting data. For example, the mass of an object determined with an analytical balance (see 9.5) should be recorded to the nearest 0.0001 g. Similarly the volume of a liquid measured with a 50 cm^3 burette should be recorded to the nearest 0.05 cm^3. (see 9.9). The same principle applies to the calculation of results. It is pointless to write down an answer directly from an electronic calculator to eight or nine significant figures when the use of a measuring cylinder for the determination of the volume at one stage in the experiment would merit a precision of only ± 2%.

The calculated result of an experiment should be reported to the number of significant figures determined by the least precise value used in the calculation.
For example,

$$\frac{0.0937 \times 25.00}{22.15} = 0.105\,756$$

would be reported as 0.1058 and not as 0.106 if it is assumed that the last figure in each value is uncertain by one digit. The result has more significant figures than the limiting (or least precise) value, i.e. 0.0937, but reporting the result as 0.106 would decrease the precision from 1 part in 937 (approx. 0.1%) to 1 part in 106 (approx. 1%).

Zeroes are not significant when they appear as the first figure or figures in a number to fix the position of the decimal point, thus 0.0937, 0.009 37 and 0.000 937 each have the same number of significant figures as 9.37 or 93.7. However, a terminal zero is significant, e.g. 0.09370 indicates an uncertainty of 1 part in 9370. Large numbers should always be expressed using index notation, e.g. 9 370 000 as 9.37×10^6, to avoid giving an impression of unattainably high precision.

(b) *Errors*
The three sources of error in a measurement are:

1. Instrumental errors, i.e. those which result from the accuracy of the measuring instrument itself.
2. The reading or observational error.
3. The setting or adjustment error.

These three types were discussed in sub-section 9.3. The written account of an experiment should include an estimate of the maximum expected error in the result. This is obtained by adding the relative errors, i.e.

$$\frac{\text{error in a measurement}}{\text{mean value of measurement}} \times 100$$

of all the quantities which appear in the final equation from which the result is calculated. The percentage error is doubled if a quantity is squared, similarly the error in any term is multiplied by n if its quantity is raised to the nth power in the equation.

(c) *A sample laboratory report*
An experiment to calibrate a Class A pipette (see table 9.3) is described as an example of the recommended method of writing up laboratory work in a practical book.

Date: *11 November 198–* *Experiment No. 7*

Experiment to calibrate a 25 cm³ pipette

Apparatus and materials A clean 25 cm^3 Class A pipette. One 100 cm^3 glass stoppered conical flask. One 0–50 °C thermometer. An analytical balance. Distilled water.

Method The pipette was rinsed out with distilled water and then filled to a point 2–3 mm above the graduation mark. The outside of the pipette was wiped with a paper tissue and the pressure of the forefinger on the top of the pipette was gently released to allow liquid to escape until the bottom of the meniscus was level with the graduation mark. The tip of the pipette was touched against the inside of the container to remove any liquid adhering to the glass. The contents of the pipette were transferred to the weighed stoppered conical flask. The jet of the pipette was held against the inner wall of the flask for 15 s after the liquid was discharged. The flask was then stoppered and reweighed. The temperature of the distilled water used to fill the pipette was noted. This procedure was repeated twice to obtain a satisfactory mean value of the capacity of the pipette.

Readings and measurements

	(1)	(2)	(3)
Mass of container + pipetteful of distilled water /g is =	92.7731	117.6937	142.6174
Mass of container /g is =	67.8479	92.7731	117.6937
Mass of distilled water /g is =	24.9252	24.9206	24.9237

Mean mass of distilled water = 24.923 g

Temperature of distilled water used for determination = 23.1 °C

Calculation

1 g of distilled water at 23.1 °C occupies 1.0034 cm^3 [see Appendix X]

Volume occupied by 24.923 g of distilled water at 23.1 °C $= 24.923 \times 1.0034 = \underline{25.008\ cm^3}$

$\therefore$ Capacity of pipette $= 25.01\ cm^3$.

Conclusion The calibration value is well within the tolerance of a Class A pipette ($\pm$ 0.03 cm^3, see table 9.3) and the nominal volume of 25.00 $\pm$ 0.03 cm^3 may be accepted.

10.2
Reporting of experimental results

All experimental data and observations should be written directly into the laboratory notebook. This includes any results which do not agree with the preconceived ideas of the experimenter. It may be that these unexpected results are caused by a fault in the instrument, for example erratic readings from a pH meter may be due to a partly discharged battery power supply or to a mistake in making up the solutions. Occasionally the data obtained provide a useful means of detecting the *cause* of the unexpected behaviour. It is also possible, of course, that the operator may be mistaken in the nature of the observations he was expecting or their magnitude. There are many examples in the history of science of important new discoveries which were made in this way. Penicillin, argon and the other noble gases, the planet Pluto, the discovery of the gas helium on the sun many years before it was found on earth were all a result of unexpected observations.

The *honest* reporting of experimental results, in spite of pressures which may be exerted from other sources, is an essential part of the training of any scientist. Little credibility is given to the results or observations of a worker once it is known that he has 'cooked' results or 'adjusted' figures to fit a graph or to agree with an expected value.

Assignment

Discuss the reasons why pressures might be brought to bear on a technician working in various types of laboratory to report experimental results dishonestly.

10.3
Presentation of experimental results

(a) *Tables*

Experimental results are usually summarised in tabulated form at first, although this is not necessarily the most appropriate form for their final presentation. Tables should always be titled and the columns headed so that it is clear what the data refer to and what units were used. Physical quantities are best tabulated as pure numbers by first dividing them by their units. For example, the following observations in an experiment to find the effect of the length (l) of a simple pendulum on its mean time of swing (t) may be listed in columns headed:

l/m	t/s	t^2/s^2
1.00	2.00	4.00
0.90	1.90	3.61
0.80	1.79	3.20
0.70	1.68	2.82
0.60	1.55	2.40
0.50	1.42	2.02
0.40	1.27	1.61
0.30	1.10	1.21
0.20	0.90	0.81

Similarly, volume, cross-sectional area, potential difference and electric current may be tabulated as volume/m^3 or cm^3, cross-sectional area/m^2 or cm^2, potential difference/volts and electric current /amperes respectively. Ideally, SI units should be used throughout, i.e. density/kg m^{-3}, pressure/Pa (or /N m^{-2}) etc.; however, experimental data should always be recorded initially in the practical notebook in the units in which they were measured. Any errors in applying conversion factors to reduce the values to SI units can then be corrected later.

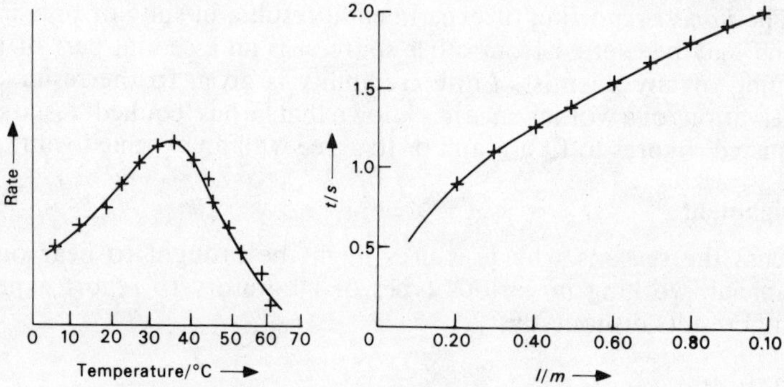

(a) Effect of temperature on the rate of the enzyme catalysed reaction of a digestive proteinase

(b) Effect of the length of a simple pendulum on its time of swing

Fig.10.1 Line graphs: enzyme catalysis and pendulum

(b) *Graphs*

Sometimes a graphical representation may be more appropriate for the final presentation of the results, especially in cases where the relationship *between the quantities* is more important than the precise magnitudes of the quantities themselves. Examples of such graphs are given in fig. 10.1 (a) and (b) which show the effect of temperature on the rate of an enzyme catalysed reaction and the effect of the length of a simple pendulum on its time of swing respectively. The axes of any graph must be labelled and, again, physical quantities should be plotted as pure numbers by first dividing them by their units. Except in special cases, such as the calibration of an instrument (see fig. 9.31, for example), the points on the graph should not be joined up by a series of short lines but a smooth curve or the best straight line should be drawn through them.

Straight line graphs The processing of data to obtain a linear relationship is an important application of the graphical method. A straight line graph indicates that the two quantities plotted are directionally proportional to one another, i.e. if one quantity is

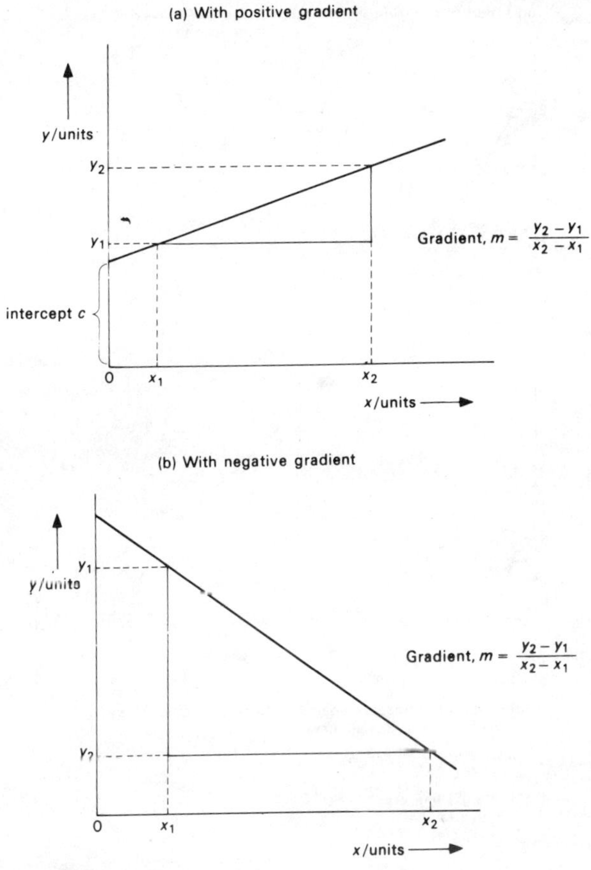

Fig.10.2 Straight line graphs: positive and negative gradients

doubled then so too is the other. This may be expressed mathematically in the form:

$$y \propto x$$

or $$y = (\text{a constant}) \times x = m \times x$$

where x and y represent the quantities and m is the proportionality content.

The equation for a straight line graph is:

$$y = mx + c$$

where y and x are the dependent and independent variables respectively and m is the gradient of the line. If $x = o$, $y = c$ and c is thus the intercept on the y-axis. The gradient m can be either positive or negative (see fig. 10.2 (a) and (b) respectively).

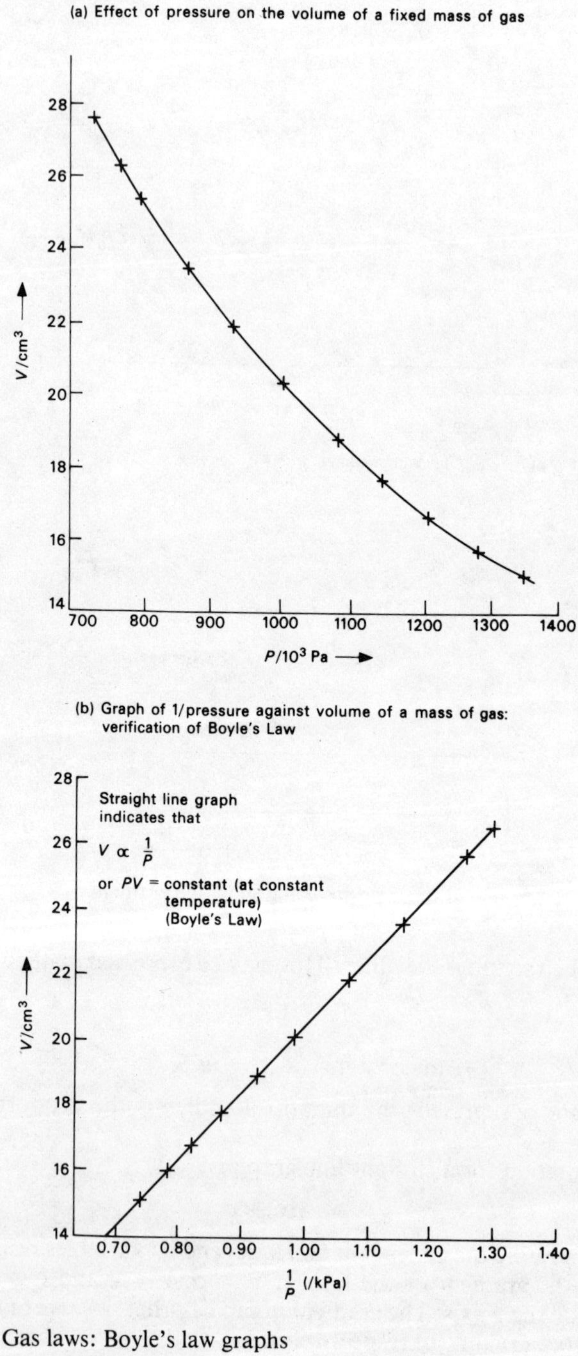

Fig.10.3 Gas laws: Boyle's law graphs

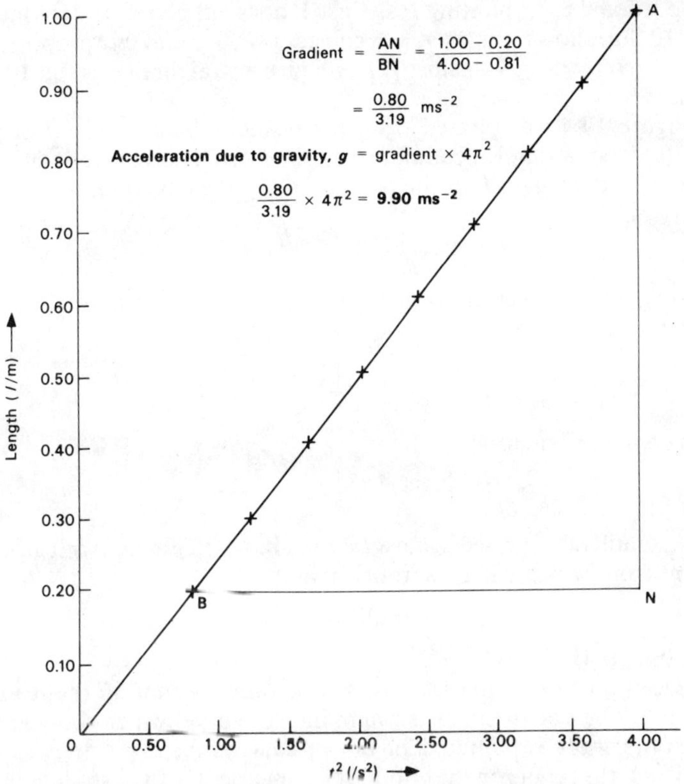

Fig.10.4 Graph of length against (time)^2 for a simple pendulum

Two examples in which results are plotted graphically to show the advantage of selecting variables which give a linear relationship are shown in figs. 10.3 and 10.4. Figure 10.3 was plotted from the results of an experiment to verify Boyle's law. *Boyle's law states that the volume of a fixed mass of gas at constant temperature is inversely proportional to its pressure*, i.e. the volume of the gas is halved if its pressure is doubled.

$$V \propto \frac{I}{P}$$

or

$$V = \frac{\text{a constant}}{P}$$

and PV = a constant (at constant temperature).

Figure 10.3(a) shows values of V plotted against the corresponding values of P. As P is not directly proportional to V, but *inversely* proportional to it, plotting P against V does not give a straight line, as fig. 10.3(a) shows. But as V = a constant/P, V is *directly* proportional to $1/P$; so plotting V against $1/P$ will give a straight line as fig. 10.3(b) indicates.

Figure 10.4 was plotted from the results tabulated in sub-section 10.3(a) from a simple pendulum experiment. The period t, of swing, is related to the length l, of the pendulum, by the equation:

$$t = 2\pi \sqrt{\left(\frac{l}{g}\right)}$$

where g is the acceleration due to gravity. As π and g are constants, it is apparent that

$$t \propto \sqrt{l}$$

or

$$t^2 \propto l$$

The complete equation is

$$t^2 = 4\pi^2 \frac{l}{g}$$

and a graph of t^2 plotted against l should be a straight line with gradient $g/4\pi^2$ from which g may be found from

$$g = \text{gradient} \times 4\pi^2$$

(see fig. 10.4).

Because of experimental errors, it is unlikely that *all* the readings taken will be exactly on the straight line (or curve) which should result when the graph is plotted. The line or curve is therefore drawn which best 'fits' the points on the graph. It is far easier to 'fit' a straight line to a set of results than to fit any other kind of curve; and any slip or misreading can readily be identified because of its departure from the

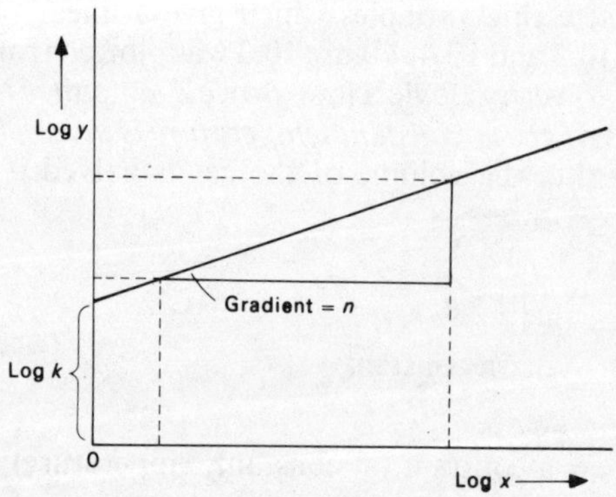

Fig.10.5 Graph of the equation $\log y = \log k + n \log x$

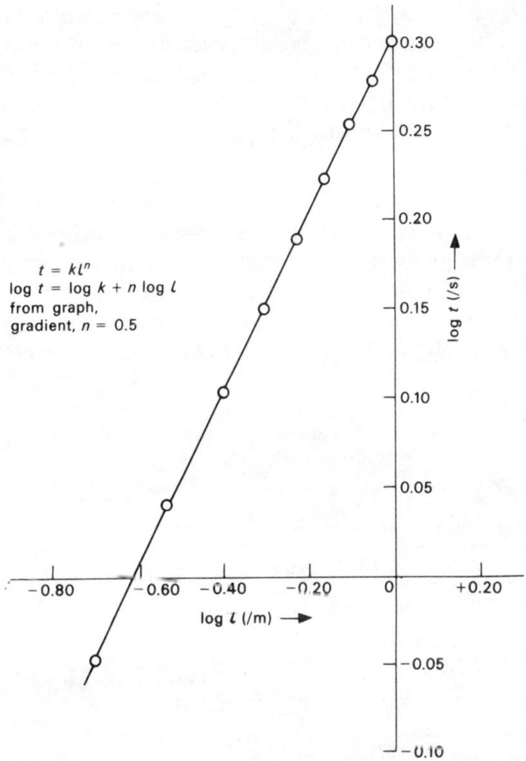

Fig.10.6 Graph of $\log t = \log k + n \log l$ for a simple pendulum

straight line, and the reading repeated. For these reasons, variables are manipulated, wherever possible, to yield a straight line graph when they are plotted.

A graph may be used to determine the relationship between two quantities, thus if

$$y = kx^n$$

the values of k and n may be found by plotting $\log x$ against $\log y$.

By taking logarithms of both sides, the equation becomes

$$\log y = \log k + n \log x$$

which has the same form as

$$y = c + mx$$

and a straight line graph is obtained of gradient n and intercept $\log k$ (fig. 10.5). The application of this technique to the simple pendulum equation is illustrated in fig. 10.6.

(c) *Other types of graph and the pictorial representation of results*
In some cases information can be conveyed or absorbed more easily if it is represented in pictorial form. The principal examples of this type of representation are the bar chart or block diagram, the histogram, the pie chart and the pictogram. Examples of these are shown in figs. 10.7–10.10.

1. *The bar chart or block diagram* consists of a series of horizontal or vertical rectangles whose length (or area) is proportional to the magnitude of the property concerned (fig. 10.7(a)). Two or three kinds of information can be compared on a single diagram by means of groups of bars which are coloured or shaded differently to aid identification (fig. 10.7(b)). Increases and decreases in a quantity may be

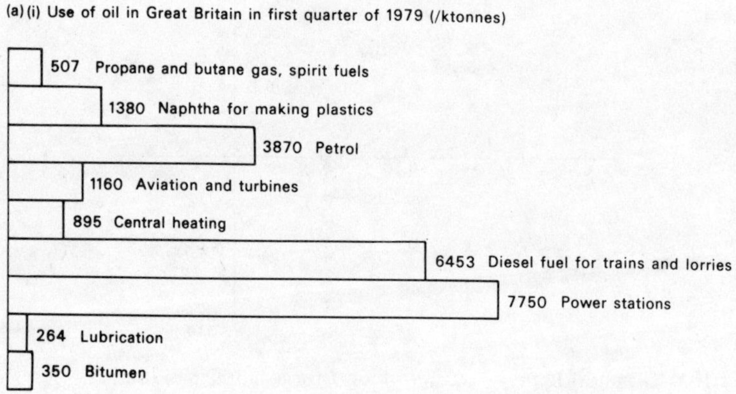

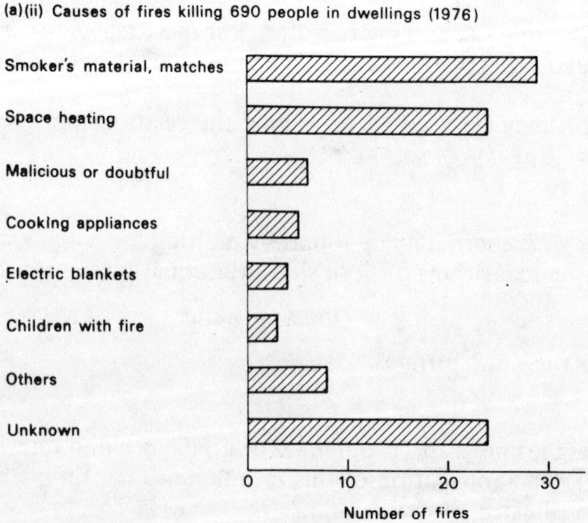

(b) Death rates from accidents in the home 1976

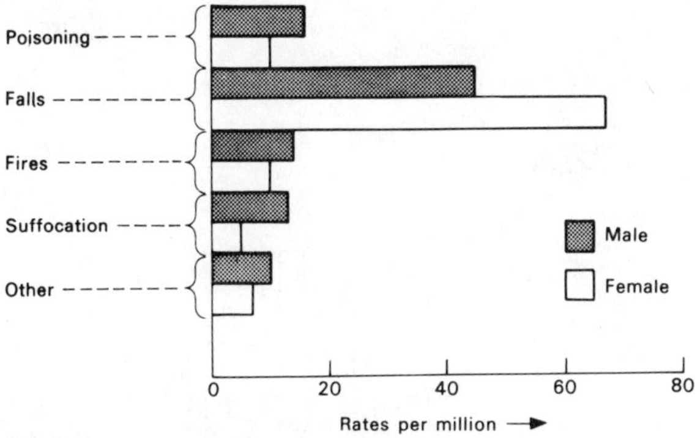

(c) Balance of payments of a firm over a ten-month period

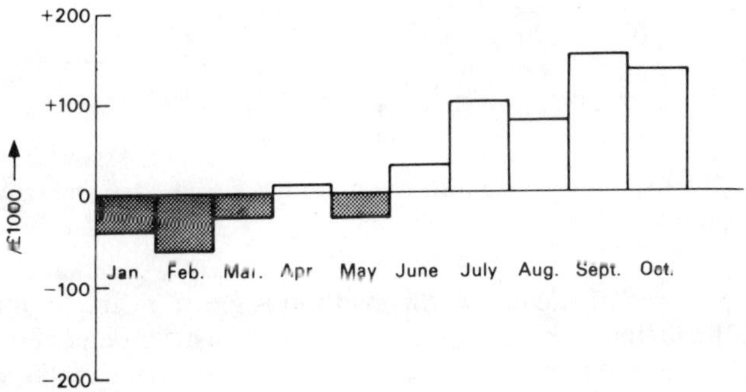

Fig. 10.7 Bar charts

shown by bars drawn in opposite directions above or below the zero line (fig. 10.7(c)).

2. *The histogram* Block diagrams in which the relative heights of the rectangles are used to represent a frequency distribution of a single variable are known as *histograms*. The example in fig. 10.8 shows the distribution of marks in an examination taken by a total of ninety-six students.

3. *The pie chart* (*or circular diagram*) is a useful method of showing how any total is divided into its sub-units. The angle subtended by any segment is proportional to the number or fraction in the category concerned. As there are 360 degrees in a complete circle, the scale is

$$360° = 100\%, \text{ or } 3.6° = 1\%.$$

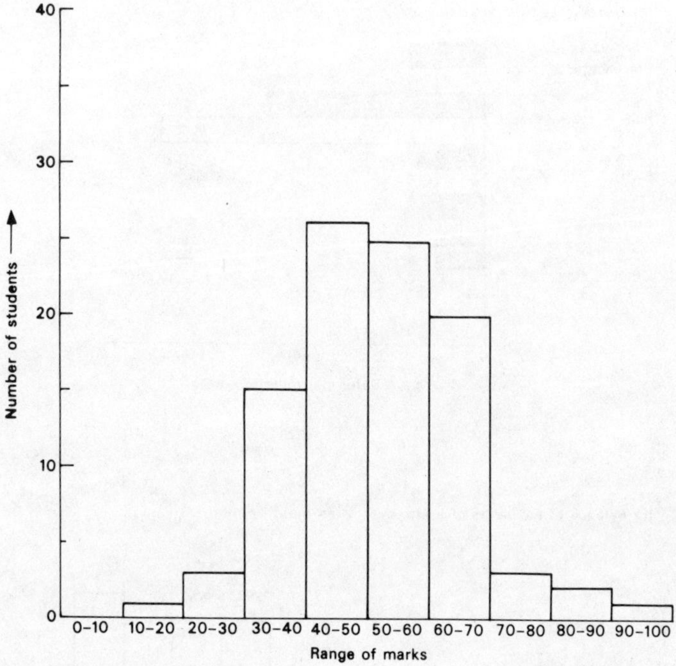

Fig.10.8 Histogram showing distribution of marks scored in an examination by ninety-six students

An example of the use of this method is given in fig. 10.9(a) which shows the different causes of accidents in industry in an average year. This information may also be conveyed in the form of a *100% bar chart* in which the relative frequencies are proportional to lengths along the bar (fig. 10.9(b)).

4. *The pictogram* This is a visual method of presenting statistical information using drawings or pictures of the subject to which the data refer. The method is restricted to the presentation of a single relationship. The difficulty of comparing objects of different sizes is overcome by using a standard symbol to represent a unit value of the data and the appropriate number of repetitions of this symbol corresponds to the magnitude of the particular quantity. The diagram obtained is thus a *pictorial bar chart*. An example is shown in fig. 10.10.

Assignment

Construct a bar chart, histogram, pie chart or pictogram to represent suitable tabulated data or statistical information from books, newspapers or surveys and reports. Why do you think the pie chart was given its name?

(a) A pie-chart: different causes of accidents in industry in a typical year

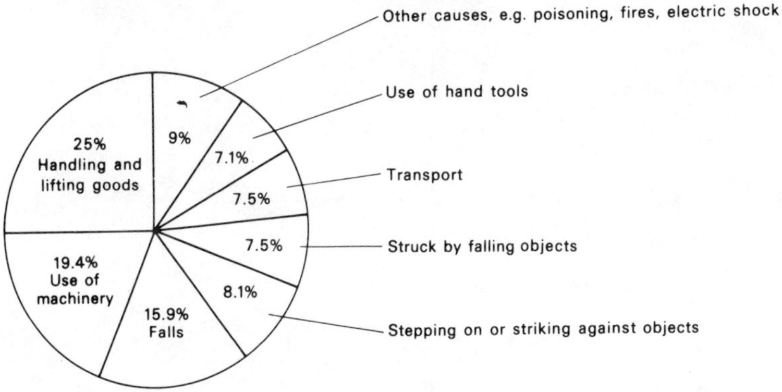

(b) A 100% bar chart: different causes of accidents in industry in a typical year

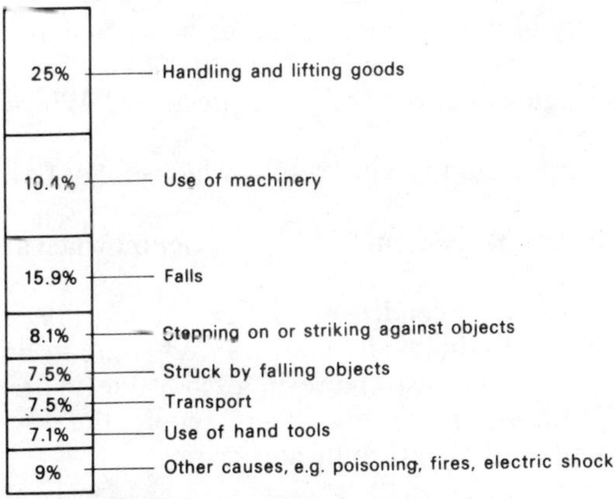

Fig.10.9 Pie chart and 100% bar chart

Questions: 10 *Scientific Reporting*

10.1 Why is it important to record *all* experimental observations in a permanent notebook?

10.2 A student weighed out 10.00 g of a substance on a top-pan balance and then transferred it to a 100 cm^3 volumetric flask. He dissolved the substance in water and made the volume of the solution up to the mark with distilled water. He then used a measuring cylinder to transfer one-third of the solution into a beaker and calculated the mass of the solute present in this sample as 3.333 33 g. Discuss.

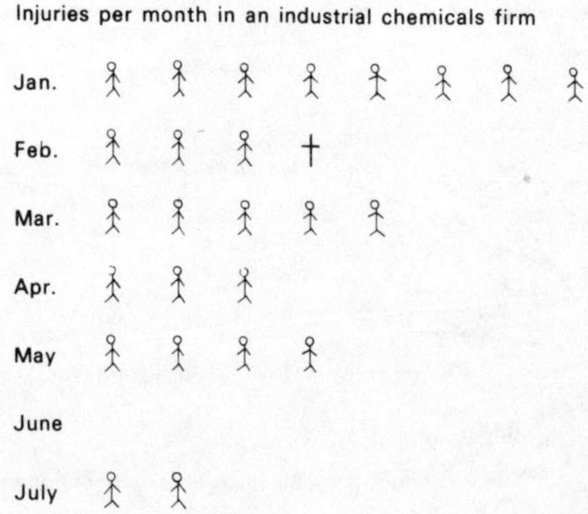

Fig.10.10 A pictogram († = a fatality)

10.3 The velocity of light in a vacuum may be written as

$$2.998 \times 10^8 \text{ ms}^{-1} \text{ or } 299\ 800\ 000 \text{ ms}^{-1}.$$

Do both these figures convey the same precision? Which method is correct and why?

10.4 Why is it important to record results which do not fit the expected trend or pattern?

10.5 What is meant by the term 'significant figure' when applied to an experimental result?

10.6 What is the recommended procedure for labelling the axes of graphs or columns of tabulated quantities? What are its advantages?

10.7 Assuming that the last figure in each value in the following calculations is uncertain by one digit, report the results to the appropriate number of significant figures:

(a) $\dfrac{5.2}{(0.306)^3}$ (b) $\dfrac{1.01 \times 10^3 \times 0.936}{312 \times 3.03}$ (c) $\dfrac{100.53 \times 1.76 \times 0.291}{(0.00372)^2}$

The answers should first be calculated using an electronic calculator.

10.8 If the uncertainty in measuring (a) the length of a simple pendulum and (b) its mean time of swing are 0.1 cm and 0.1 s, respectively, calculate the percentage relative errors in each of the following measurements:

l/m	t/s
1.00	2.00
0.10	0.63

What is the maximum expected error in each case in the acceleration g due to gravity when the result is calculated from the equation:

$$g = 4\pi^2 \frac{l}{t^2}?$$

10.9 What does a straight line graph indicate?
10.10 Describe how you would determine the values of the constants k and n in the equation

$$z = kv^n$$

from measured values of z and v.
10.11 What are bar charts, histograms, pie charts and pictograms? What are they used for?

E Laboratory skills

Section 11: *The expected learning outcome of this section is that the student should be able to care for the items of equipment common to all laboratories*

Specific objectives: *The expected learning outcome is that the student should be able to:*

11.1 Clean glassware using: water, a commercial product, a chromic acid technique.

11.2 Select the appropriate method (of those in 11.1) in a specified situation.

11.3 Maintain a record of breakages and report them periodically as directed.

11.4 State the electrolyte used in common secondary cells.

11.5 Charge common laboratory batteries.

11.1
Cleaning glassware

The extent to which confidence in the reliability of experimental work is dependent on the availability of clean glassware is obvious, but not generally recognised. A chemist, for example, would have little faith in the accuracy of a titration reading if the solutions were prepared in contaminated volumetric flasks and the result was obtained using a burette smeared with grease and a dirty pipette. Similarly, anyone working in a biological or physical laboratory would express concern if there were a possibility that the glassware he (or she) was using was still contaminated with a culture or by the radioactive waste of a previous experiment. The methods used for the sterilisation of glassware used in microbiological experiments and for the destruction or disposal of pathogens were described in sub-section 4.6. The general methods of cleaning laboratory glassware are described in this section.

The principal cleaning agents are:

1. Water.
2. Commercial products, e.g. 'Teepol', or specialised laboratory detergents such as 'Quadralene' or 'Decon 90'.
3. Chromic acid. This is prepared by mixing sodium dichromate (VI) (50 g) with 75 cm^3 of water and then adding 400 cm^3 of concentrated sulphuric (VI) acid. (*Care*: A face shield must be worn and the acid should be added a few cm^3 at a time as the mixture is stirred. The deep

reddish-brown solution should be treated with care and should not be allowed to come into contact with the skin.)

4. Organic solvents, such as propanone (acetone) or tetrachloromethane (carbon tetrachloride), are not recommended for general use, but are required in some cases to dissolve grease and other organic solvents.

5. Alkalis, e.g. 5% aqueous solution of sodium hydroxide (caustic soda) or potassium hydroxide (caustic potash).

6. Aqueous 1% potassium manganate (VII).

11.2
Selection of appropriate cleaning method

Most glassware in normal laboratory use may be cleaned by washing in water containing a little 'Teepol' or other detergent. It should then be rinsed well under a running tap and finally washed with a little distilled water and/or a small amount of propanone (acetone) and placed in a cabinet to dry. If the stains do not respond to this treatment the glassware should be immersed in a solution of an oxidising agent, such as chromic acid or aqueous 1% potassium manganate (VII). (*Care*: A face shield or goggles and rubber gloves must be worn when using chromic acid.) These solutions are particularly useful for volumetric glassware and flasks or burettes can be filled with the solution and left overnight. Small pieces of glassware may be placed in a bowl of the solution or—in the case of pipettes—in a tall measuring cylinder. Any brown stains remaining after treatment with potassium manganate (VII) solution may be removed with a little concentrated hydrochloric acid.

Traces of grease or oil may be removed by rinsing with a small amount of propanone (acetone). (*Care*: This solvent is highly inflammable.) Heavier deposits of tarry substances, resins, gum, oil or other organic residues may require specialised laboratory detergents. These products are frequently used in ultrasonic cleaning tanks which increase the speed and efficiency of the cleaning process. In each case the glassware is rinsed thoroughly first with tap water and then with a little distilled water before draining and drying.

In general, solutions of alkalis (e.g. aqueous sodium hydroxide or potassium hydroxide) should not be used for cleaning volumetric glassware. All stoppers and taps must be removed before placing the glassware in the alkali, otherwise they may become fused in place and the apparatus should be rinsed thoroughly with water after cleaning.

Assignment

What methods are used to clean the glassware in the laboratory where you work? Compare these methods with those employed by other students in the class who work in different types of laboratory.

11.3
Breakages

Breakages occur from time to time in even the best managed laboratories. This applies not only to glassware, but also to meters, rheostats, stopclocks and to any other piece of laboratory equipment. A large proportion of these incidents are avoidable and need not have occurred with a little forethought. For example, thin-walled 'Pyrex' glass may be heated directly in a bunsen burner flame or dropped into boiling water without harm, but it is not a transparent form of stainless steel and the fact that a pipette or filter funnel placed flat on the bench close to the edge rolls off and smashes on to the floor should not be unexpected. Similarly, balances, rheostats and galvanometers must not be overloaded nor should other equipment be used in a manner contrary to the manufacturer's instructions.

Glassware breakages can be a rarity instead of a commonplace occurrence in a laboratory, provided that a certain minimum care is taken. Glass apparatus must be properly clamped and supported as the distillation equipment shown in fig. 11.1 illustrates. Large bottles filled with liquids should not be picked up with wet hands and apparatus, bottles, flasks, thermometers etc. should not be placed close to the edges of shelves or benches where they can be knocked off on to the floor.

All breakages, except those of such minor items as test tubes and microscope slides, should be recorded in a book and the equipment repaired (if this is economic) as soon as possible. A regular check should be kept of all breakages as this is the major factor in depleting stock in many laboratories. They should be reported periodically, indicating the present stock and any items which are in short supply so that replacements can be ordered. An occasional 'clampdown' to reduce the occurrence of careless breakages is also recommended.

Assignment

1. What is the procedure for reporting breakages in your laboratory?
2. What procedure is followed for the repair of damaged equipment?
3. Discuss the reasons for a number of recent laboratory breakages and describe how these could have been avoided.

11.4
Secondary cells

An *electrochemical cell* converts the chemical energy of the substances stored within it into electrical energy. Cells are of two types: *primary cells* and *secondary cells*. A primary cell is one in which the active substances are consumed as the cell supplies an electric current and are thus gradually used up. Once these materials are depleted the voltage drops

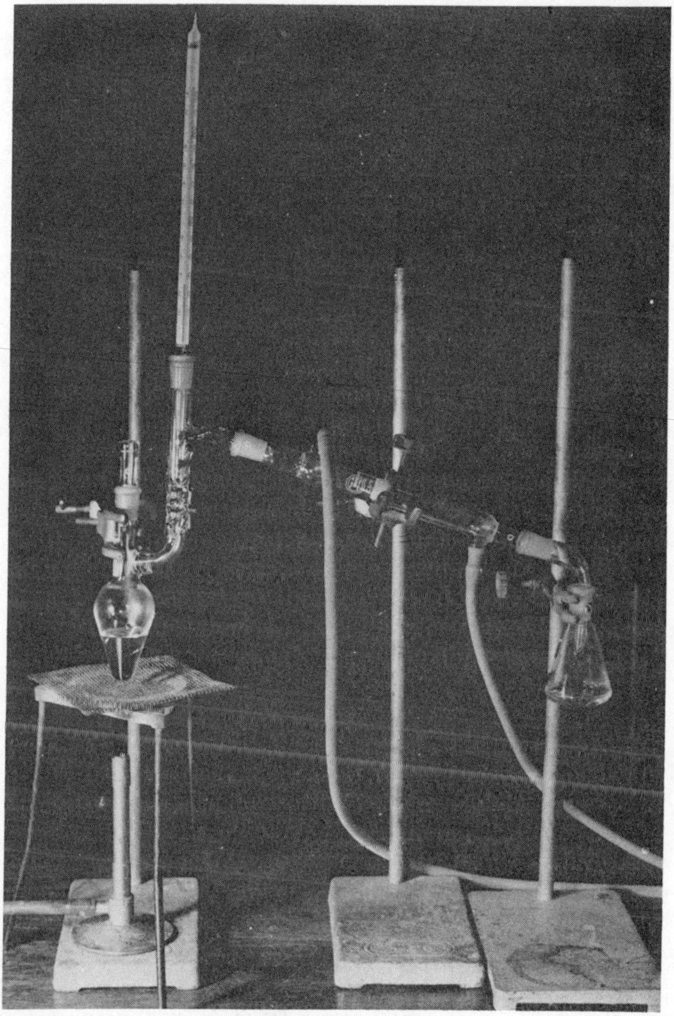

Fig.11.1 Clamping of glass distillation equipment

and unless the chemicals can be replaced the cell has to be thrown away. The dry batteries used in torches, radios and portable cassette recorders etc. or the Mallory–Ruben mercury cell (fig. 11.2) used in electronic watches, cameras and small calculators are examples of primary cells.

A secondary cell is one in which the chemical reaction which supplies the current can be reversed. This is done by passing a current *into* the cell from an external source in the opposite direction to that in which the current is supplied by the cell (fig. 11.3). Secondary cells thus act as

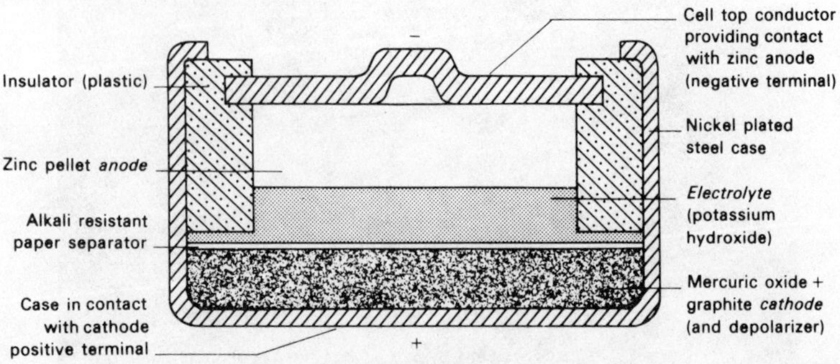

Fig.11.2 Mallory-Ruben mercury cell

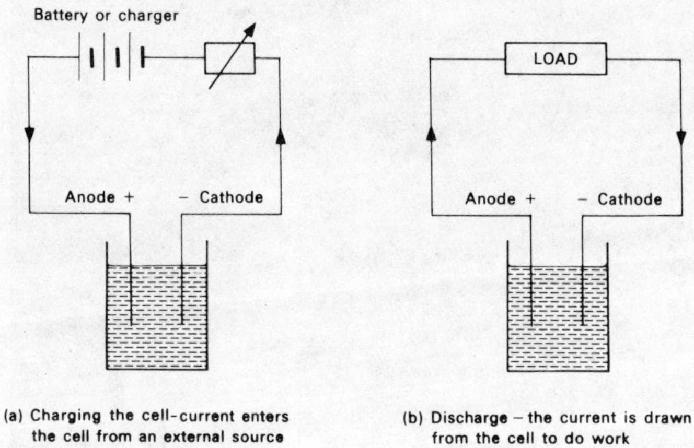

Fig.11.3 Charge and discharge of secondary cells

storage batteries for electrical power and operate by a charge–discharge process. The car battery is an example of a secondary cell. The large amount of electrical energy withdrawn from the battery in starting the car is quickly restored once the car is moving by the current generated by the dynamo or alternator. The current entering the cell reverses the chemical reactions which occurred during discharge and thus recharges the battery.

The commonest secondary cells in regular use in laboratories are the lead–acid accumulator or battery with an electromotive force (e.m.f.) of about 2 V per cell and the nickel–iron (NiFe) or nickel–cadmium alkaline cells with an e.m.f. of about 1.2 V per cell. The electrolytes in these two types of cell, i.e. the acid cell and the alkaline cell, are aqueous sulphuric (VI) acid and aqueous potassium hydroxide respectively (see table 11.1).

Table 11.1 *The electrolytes used in secondary cells*

Cell	e.m.f /V	Electrolyte
Lead–acid battery	2	Aqueous sulphuric (VI) acid (density 1.25 g cm^{-3})*
Nickel–iron (NiFe) alkaline cell	1.2	Aqueous potassium hydroxide
Nickel–cadmium alkaline cell	1.2	Aqueous potassium hydroxide

Note: Both these electrolytes are corrosive. They are harmful to the skin and eyes.

* Battery acid may be prepared as follows: Add 220 cm^3 of concentrated sulphuric (VI) acid carefully to 750 cm^3 of cold, distilled water. (*Care*: A face shield must be worn and the acid should be added 20–30 cm^3 at a time to the water to moderate the temperature increase and the mixture should be stirred continually.) Make the volume of the solution up to one litre with distilled water and stir well. Check the specific gravity of the mixture with a hydrometer (see fig. 11.4) and adjust to 1.250 by adding more distilled water or concentrated sulphuric (VI) acid respectively if the reading is high or low respectively.

The capacity of a secondary cell is usually given in ampere hours (A h), and is a measure of the quantity of electricity it can supply when the

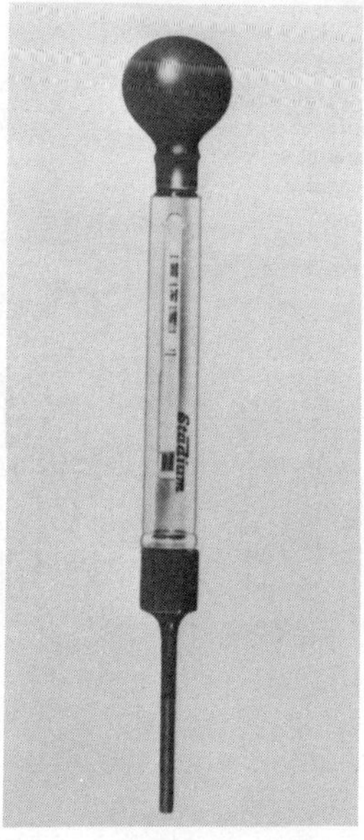

Fig.11.4 A hydrometer

fully charged cell is discharged at a uniform rate. Provided the discharge current does not exceed the current rating of the battery, a 60 A h cell, for example, will supply a current of 3 A for 20 h or 5 A for 12 h etc.

$$1\ \text{A h} = 1\ \text{A} \times (60 \times 60)\ \text{s} = 3600\ \text{C}$$
$$1\ \text{coulomb (C)} = 1\ \text{ampere second.}$$

The ampere hour is not an SI unit; nevertheless, it does provide a convenient and widely employed measure of cell capacity.

(a) *The lead–acid secondary cell*

The positive plate of the lead–acid cell consists of a framework made of a lead–antimony alloy (containing 94% lead and 6% antimony) supporting a paste of lead(IV)oxide (PbO_2) which is the active ingredient. The negative plate has a similar structure, but the framework is packed with a soft, spongy form of lead. A number of alternating positive and negative plates are mounted close to one another to provide a large surface area and to decrease the internal resistance of the cell (fig. 11.5). Separators of glass, plastic or other insulating materials are placed

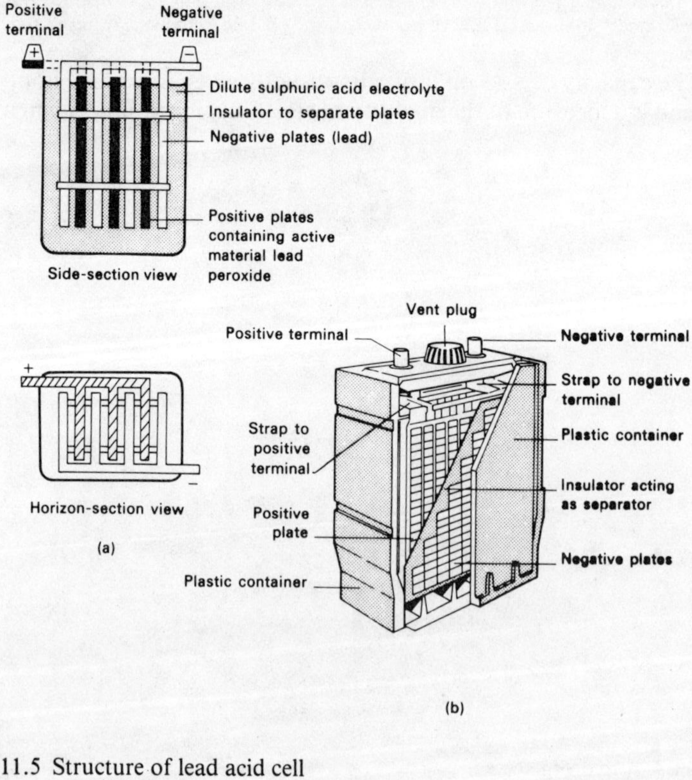

Fig.11.5 Structure of lead acid cell

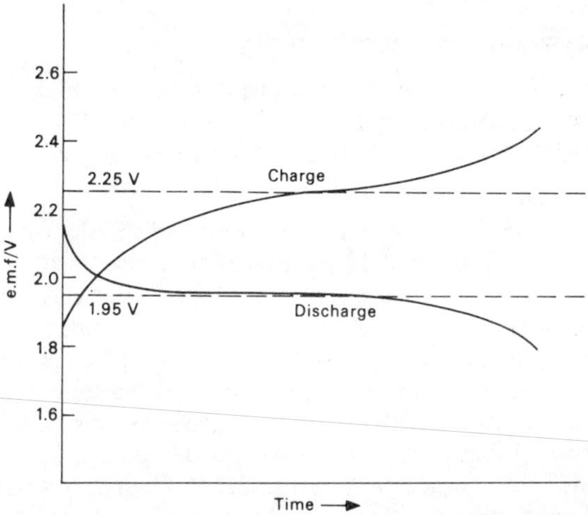

Fig.11.6 Variation of e.m.f. of lead-acid cell

between the plates to prevent them touching one another. The container is made of a hard, durable plastic material.

The e.m.f. of a fully charged lead-acid cell is about 2.1 V. This slowly drops on discharge (fig. 11.6) and should not be allowed to fall below 1.8 V before recharging. The withdrawal of an electric current from the cell gradually converts the lead dioxide on the positive plate into lead sulphate and water, thus diluting the sulphuric acid electrolyte and causing its specific gravity (or relative density) to drop from about 1.26 to approximately 1.17. Lead sulphate is also formed on the cathode during discharge. The chemical equations for these reactions are given in table 11.2.

(b) *The nickel–cadmium alkaline cell*

The positive plate of this cell consists of a paste of active hydrated nickel (III) oxide ('nickelic hydroxide') surrounded by strips of nickel on a nickel plated steel support. The cathode is a layer of spongy cadmium on a nickel plated steel support. In the nickel–iron cell (or NiFe cell, after the corresponding chemical symbols of the two metals) this spongy cadmium is replaced by finely divided iron containing a small amount of mercury. The nickel–cadmium cell supplies higher discharge currents and operates better at lower temperatures than the nickel–iron cell. Accumulators of alkaline cells are lighter in weight per kilowatt hour of energy capacity than lead–acid accumulators and are more robust, but cell for cell have a lower e.m.f.

11.5
Charging common laboratory batteries

The chemical reactions in the lead–acid or nickel–iron and nickel–cadmium alkaline cells are reversed during charging (see table 11.2). The electrons are provided by the external power source.

(a) *The lead–acid cell*
The current used to recharge the lead–acid cell should not be too high (nor should the cell be 'shorted') as this might cause the plates to buckle or displace the active substances and damage the cell. Most trickle chargers provide a current of 2–5 A, which is adequate for overnight charging of a partially discharged battery, although occasionally boost chargers capable of providing a higher current for a short period may be useful. The relative density of the sulphuric (VI) acid electrolyte increases during charging to a maximum of about 1.26 at which the e.m.f. of the fully charged cell is about 2.2 V. Hydrogen and oxygen are freely evolved from the negative and positive plates respectively when the battery is fully charged. This is known as 'gassing'. Cells must never be charged close to a naked flame as this gaseous mixture is highly explosive.

On standing, lead–acid cells undergo a slow self discharge at a rate of about 1 or 2 per cent per day owing to the reaction of the electrodes with impurities such as iron. This self-discharge rate increases rapidly if the battery is topped up with tap–water or if contaminated distilled or deionised water is used. A battery should first be fully charged and then drained of its electrolyte if it is to be stored for a long time.

(b) *Alkaline cells*
Nickel–iron or nickel–cadmium alkaline batteries have an e.m.f. of about 1.2 V per cell. They are more expensive than the lead–acid cell, but—unlike the lead-acid battery—they can be inadvertently short circuited, overcharged or left uncharged for long periods without major deterioration. The electrolyte should be changed if its specific gravity falls below 1.160. When this occurs the cell should be fully discharged at the normal rate to about 0.80 V per cell. The cells should then be drained completely and immediately refilled with new electrolyte (see table 11.1) until the plates are covered to the correct depth. The cells are then charged for 15 h at the normal 7 h charge rate or proportionately longer with a lower charging current and the specific gravity of the aqueous potassium hydroxide electrolyte checked to see that it is about 1.190. The level of the electrolyte should be checked regularly when in normal use and topped up with distilled water to replace evaporation losses. Sulphuric acid or battery acid must NOT be used for this purpose as it will damage the cell.

Assignment

Examine the secondary cells used in your laboratory. Check the

Table 11.2 *Chemical reactions for charge and discharge of secondary cells*

Cell	e.m.f. /V	Electrodes for discharge (+)	Electrodes for discharge (−)	Electrolyte	Chemical reactions (−) electrode	Chemical reactions (+) electrode
Lead–acid	2.0	PbO_2	Pb	$H_2SO_{4(aq)}$	$PbO_2 + SO_4^{2-} + 4H^+ + 2e^- \rightleftharpoons PbSO_4 + 2H_2O$	$Pb + SO_4^{2-}{}_{(aq)} \rightleftharpoons PbSO_{4(s)} + 2e^-$
Nickel–iron (NiFe cell)	1.37	'$Ni(OH)_3$'	Fe	$KOH_{(aq)}$		$Fe + 20H^- \rightleftharpoons Fe(OH)_2 + 2e^-$
					$Ni(OH)_3 + e^- \rightleftharpoons Ni(OH)_2 + OH^-$	
Nickel–cadmium	1.30	'$Ni(OH)_3$'	Cd	$KOH_{(aq)}$		$Cd + 2OH^- \rightleftharpoons Cd(OH)_2 + 2e^-$

electrolyte levels and top up with distilled water if required. Check the specific gravity of charged and uncharged cells with a hydrometer. (*Care*: the electrolytes are harmful to the skin and eyes.)

Questions: 11 *Laboratory skills*

11.1 Why is it essential to wear goggles or a face shield when using or preparing chromic acid?
11.2 Name *three* different cleaning agents for laboratory glassware and give examples of their use.
11.3 Why is it essential to remove taps and stoppers before soaking glass apparatus in an aqueous solution of sodium hydroxide (caustic soda)?
11.4 Potassium manganate (VII) solution often leaves a brown stain on glassware. How may this stain be removed?
11.5 Why is it important to maintain a record of breakages and to report them periodically?
11.6 Give *three* methods of reducing the number of laboratory glassware breakages.
11.7 What is an electrochemical cell?
11.8 Distinguish between a primary cell and a secondary cell. Give *two* examples of each.
11.9 Name the electrolyte in (a) a lead–acid battery and (b) an alkaline cell.
11.10 Why is it possible to determine the state of charge of a lead–acid battery using a hydrometer?
11.11 What are the advantages and disadvantages of a dry cell (e.g. a battery for a torch or portable radio) and a lead–acid battery?
11.12 Give *two* advantages and *two* disadvantages of the lead–acid battery compared with the alkaline cell.
11.13 Why is it dangerous to light a cigarette while charging a lead–acid battery?
11.14 Why is it necessary to add distilled water to lead–acid cells from time to time? Why isn't tap water used for this purpose?
11.15 What is the reason for recharging a lead–acid battery every few weeks if it is not being used?

F First-aid

Section 12: *The expected learning outcome of this section is that the student should be able to deal, in the first instance, with a variety of laboratory accidents*

Specific objectives: *The expected learning outcome is that the student:*

12.1 *Recognises the necessity of careful siting of first-aid boxes, eye-wash bottles, and first-aid points.*
12.2 *Describes the desirable minimum contents of a first-aid box for laboratory use.*
12.3 *States the priorities in administering first-aid.*
12.4 *States the purposes of artificial respiration.*
12.5 *Demonstrates a knowledge of one method of artificial respiration.*
12.6 *Describes the use of first-aid dressings for surface wounds.*
12.7 *States the methods which could be used for controlling bleeding from a wound.*
12.8 *States the locations of the main pressure points.*
12.9 *Demonstrates the use of a pressure point.*
12.10 *Lists a set of procedures for general treatment of poisons.*
12.11 *Describes the procedures for the first-aid treatment of thermal burns.*
12.12 *Describes the procedures for the first-aid treatment of chemical burns.*
12.13 *Lists the common antidotes for chemical burns by acids, alkalis, phenol, bromine and phosphorus.*
12.14 *Demonstrates a correct procedure for the use of eye-wash bottles on self and on another person.*
12.15 *Lists a sequence of actions to be taken on finding a casualty believed to be suffering from electric shock.*
12.16 *Describes the first-aid treatment of wound shock.*
12.17 *Describes the procedure to be adopted for an unconscious casualty.*
12.18 *Recognises the need for immobilising a casualty who has suffered a possible fracture.*
12.19 *Recognises the need for a reporting procedure in regard to accidents in the laboratory.*
12.20 *Describes such a reporting procedure.*

Definition of first-aid

First-aid may be defined as the immediate treatment and care given to the victim of an accident until the services of a doctor or other qualified medical practitioner can be obtained.

It should be remembered that this treatment is only temporary and is given to achieve the following three objectives:

1. to sustain life,
2. to prevent the victim's condition from becoming worse and
3. to promote the victim's recovery.

Any laboratory is potentially a dangerous place to work in, but fortunately serious injuries are rare. Nevertheless, the danger exists and any technician should know how to deal with cuts, burns and other common (and frequently avoidable) injuries, as well as providing immediate, safe treatment as required in the event of a more serious accident.

The methods of treating the injuries which result from the common laboratory accidents listed in table 12.1 are described in the following sub-sections. However it should be remembered that first-aid is a skill which can be learned only by proper training and practice. Attendance at one of the courses organised by the British Red Cross Society or St John's Ambulance Brigade which leads to the award of a First-aid Proficiency Certificate is therefore recommended. These certificates are valid for three years, but may be renewed after a refresher course and re-examination to ensure that the first-aider is proficient and has kept his knowledge and skill up to date.

Table 12.1 *Common injuries resulting from laboratory accidents*

Condition or injury	Nature of accident	Particular hazard	Action/ Treatment
Asphyxia	Exposure to fumes or poisonous gases. Electric shock, suffocation etc.	Oxygen deficiency can cause brain damage or death.	Remove victim from smoke-filled room etc. Clear air passages and apply artificial respiration (see sub-section 12.5) if necessary.
Cuts and scratches	Cuts from broken glassware, knives, scalpels, etc.	Heavy blood loss, shock, and possibly death.	Control bleeding (see sub-sections 12.7–12.9) and treat for shock (see sub-section 12.16).
	Lacerations from falls, moving	Blood loss and possibility of	Control bleeding (12.7–12.9). If

Table 12.1 *(Continued)*

Condition or injury	Nature of accident	Particular hazard	Action/ Treatment
	machinery, barbed wire etc.	infection.	serious, cover wound with sterile dressing and seek medical attention. Minor injuries should be washed thoroughly and covered with a sterile dressing (see sub-section 12.6).
Burns and scalds	Fires or touching hot objects, steam, boiling or hot liquids, corrosive substances.	Painful scarring, tissue damage, shock and death.	Immediately cool area of burn and then cover with sterile dressing, (see sub-section 12.11). Treat for shock (see sub-section 12.16). Transport severely burned casualties to hospital urgently.
Clothing on fire		Severe burns and shock	Lay the victim on the floor and immediately roll in blanket or coat to extinguish flames (see sub-section 12.11). Treat for burns and shock.
Uncon-sciousness	Asphyxia, shock, head injuries, faint-ing, poisoning.		See sub-section 12.17
Poisoning	Escape of poisonous fumes, pipetting dangerous substances by mouth (see 9.8) or by absorption through the skin.	Damage to lungs, throat, stomach and internal organs and possibly death.	See sub-section 12.10.
Splashes of chemical in the eye		Eye damage and loss of sight.	Wash thoroughly (see sub-section 12.14) and obtain urgent medical attention.

Table 12.1 *(Continued)*

Condition or injury	Nature of accident	Particular hazard	Action/ Treatment
Splashes of chemicals on the skin		Chemical burns, poisoning by skin absorption, dermatitis.	Wash thoroughly and neutralise if possible (see sub-sections 12.12 and 12.13).
Fractures/ dislocations	Falls or heavy objects toppling onto victim.	Internal injuries. Further damage if the affected part is moved.	Immobilise victim (see sub-section 12.18) and summon medical attention.
Electric shock	Contact with live leads or equipment	Burns. The casualty may stop breathing and his heart may cease beating (see sub-section 1.8). Severe shock. Fractures, cuts and other secondary injuries from falls etc. caused by the shock.	Switch off electricity supply. Treat burns (see sub-section 12.11). Apply artificial respiration and heart massage if required (see sub-sections 12.5 (b) and (c)). Treat for shock (see sub-section 12.16). Treat secondary injuries (see sub-section 12.15). Obtain medical treatment.
Strain, sprain, back injuries, hernia (rupture)	Falls, attempting to lift loads incorrectly, (see sub-section 7.7).		Refer to doctor or hospital.

12.1
Siting of first-aid boxes, eye-wash bottles and first-aid points

All first-aid equipment should be placed at or close to the sites where accidents are likely to occur. First-aid boxes and eye-wash bottles should be readily accessible and fitted where they can be reached conveniently even by the smallest person in the laboratory without having to stand on a chair or step ladder. They should be placed where they can easily be seen. The green colour coding (see section 6) is a useful guide, but everyone working in a laboratory (including cleaners, porters, typists and other non-scientific staff) should know the exact whereabouts of all safety equipment and emergency exits. They should also know how to use this equipment, i.e. first-aid boxes, eye-wash bottles, fire extinguishers, fire blankets, etc.

First-aid boxes should never be locked, nor should equipment such as scissors or adhesive plaster be removed and used for other purposes. In large laboratories it may be desirable to have more than one eye-wash bottle and first-aid box.

Assignment

Note the location of the first-aid boxes, eye-wash bottles and other first-aid equipment in (1) the college laboratories and (2) at work. Comment on the suitability of the site chosen and on the number of first-aid boxes and eye-wash bottles provided.

12.2
Minimum contents of a first-aid box for laboratory use

First-aid boxes should be kept as simple as possible. The exact contents are dependent on the type of laboratory and on the nature of the

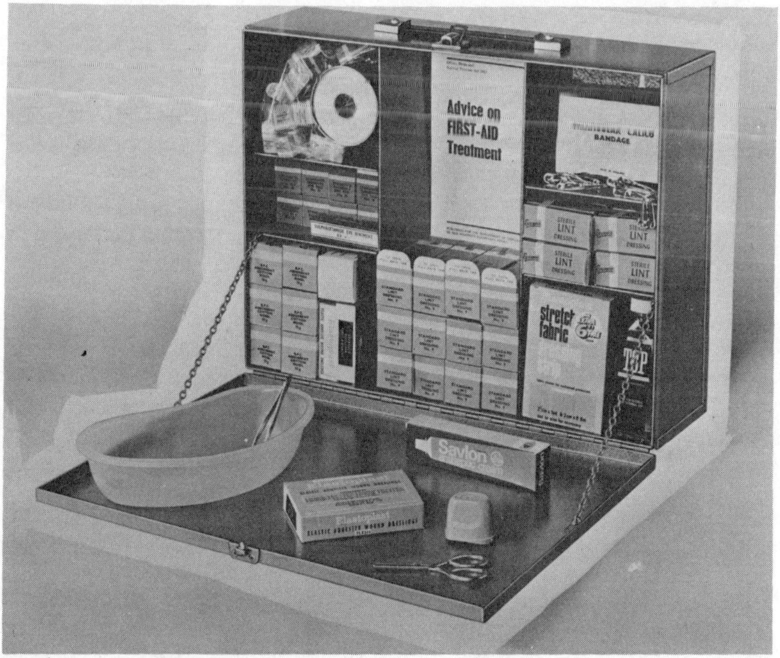

Fig.12.1 A first-aid box

hazards involved. For example, it would be foolish not to have the antidote or immediate treatment available in a laboratory where people are working with cyanides or hydrofluoric acid, but in a school, hospital or physical laboratory these facilities would not be required. The suggested minimum contents of a first-aid box (fig. 12.1) are:

one pair of scissors with blunt ends (but *not* blunt blades—scissors are no good if they do not cut);
bandages—an assortment of different sizes;
one triangular bandage for use as a sling;
one tin of adhesive plasters and dressings;
one rubber bandage or pressure bandage;
sterilized gauze and large dressings;
sterilized cotton wool;
sterilized eyepads in separate sealed packets;
small and medium sized sterilized unmedicated dressings for injured fingers, hands etc.;
one bottle of 'TCP' or other mild antiseptic solution;
safety pins;
small forceps;
one clinical thermometer;
a copy of a leaflet giving advice on first-aid treatment (e.g. Form 1008 which is supplied by the Health and Safety Executive and is available from the HMSO);
the telephone numbers of the nearest doctor and ambulance (see Appendix XI)

This list of contents is based on the legal requirements for industrial laboratories which form part of a factory. The requirements for a school, a college or another laboratory may differ slightly from this. Additional items may be provided on the advice of a firm's medical adviser, for example in many laboratories a stretcher or oxygen respirator may also be required. It is also useful to have a cup, spoon and a blanket stored close to the first-aid box.

One person in each firm or laboratory should be responsible for checking the contents of the first-aid boxes at regular intervals and for replacing any items which may have been used. It is also worth while to keep a small notebook and pencil in the first-aid box to record all accidents with a brief note of the treatment given, although this would be additional to the accident reporting discussed in sub-sections 12.19 and 12.20.

Assignment

List the contents of the first-aid box. Are the items adequate and available in sufficient number to deal with the types of accident which could be expected from the nature of the work carried out in your laboratory?

12.3
Priorities in administering first-aid

The order of priorities to be adopted when administering first aid is decided by its principal objective—to save life. Equally important, nothing should be done which endangers the lives of others. It will not help the victim if you are killed while attempting to rescue him from an electrical fault or from a room filled with smoke or poisonous fumes. It is essential for the first-aider to keep calm and to quickly assess the situation. Time is important and in serious accidents the first two or three minutes can make the difference between life and death. Precious seconds must not be wasted. The precise sequence of actions is governed by the circumstances, but the following order is of general application:

1. Quickly separate the victim from the hazard (*provided it is safe to do so*).
2. Ensure that the patient's breathing is maintained. If the victim isn't breathing begin artificial respiration immediately (see 12.4 and 12.5).
3. Control serious bleeding to prevent heavy blood loss (see 12.7).
4. If the victim is unconscious place him in the recovery position (see fig. 12.2 and sub-section 12.17). This eliminates the possibility of asphyxiation should the patient vomit or his tongue fall back and block the airway to his lungs.
5. Prevent shock (see 12.6).
6. Treat burns (see 12.11) and deal with localized injuries, such as cuts and grit or other foreign bodies in the eye.

Once these life-preserving steps have been taken the first-aider should reassure the patient and remain calm and confident. Of course, if other people are present their assistance should be sought to telephone for an ambulance if the injury is sufficiently serious to merit hospital treatment or to summon the fire brigade or other services which may be required. Always give clear instructions about the number to phone, with details of the nature and precise location of the accident, the number of casualties and the type and seriousness of the injuries. They can then help to care for the victims until skilled medical assistance arrives or deal with the cause of the injury, i.e. with the spilled chemicals, broken glassware etc. or fire (provided it is only a

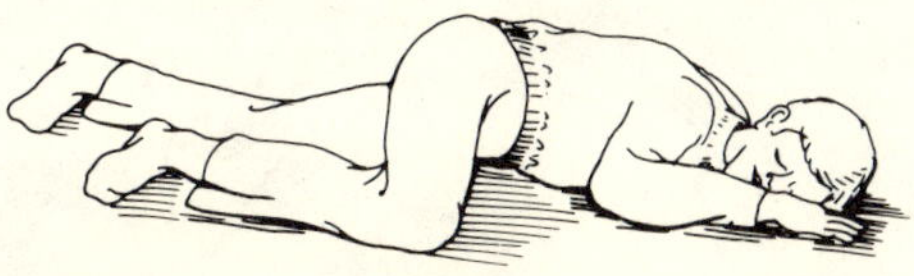

Fig.12.2 The recovery position

small one). If the casualty is taken to hospital make sure that his (or her) parents or family are tactfully informed.

12.4
The purposes of artificial respiration

(a) *Introduction: The respiratory and circulation systems*
For the purposes of first-aid, respiration (or breathing) may be defined as the process by which oxygen passes from the air into the blood, while carbon dioxide (a waste product of respiration) is expelled from the blood into the air. The atmosphere contains about 20% by volume of oxygen and 79% of nitrogen. Expired air still contains about 16% of oxygen and this high concentration explains the effectiveness of the expired air or mouth-to-mouth method of artificial respiration (see 12.5).

The gaseous exchange takes place in the *lungs*, which occupy the greater part of the chest (or thoracic cavity) and are situated on either side of the heart (fig. 12.3). The *trachea* (or windpipe) is divided into two branches, the *right* and *left bronchus*, which then pass into the lungs. These branches then subdivide into a large number of fine tubes known as *bronchioles* which divide further and lead to the many tiny air-sacs (*alveoli*) in the lungs. The alveoli are surrounded by a fine network of capillary blood vessels through which the exchange of oxygen and carbon dioxide takes place.

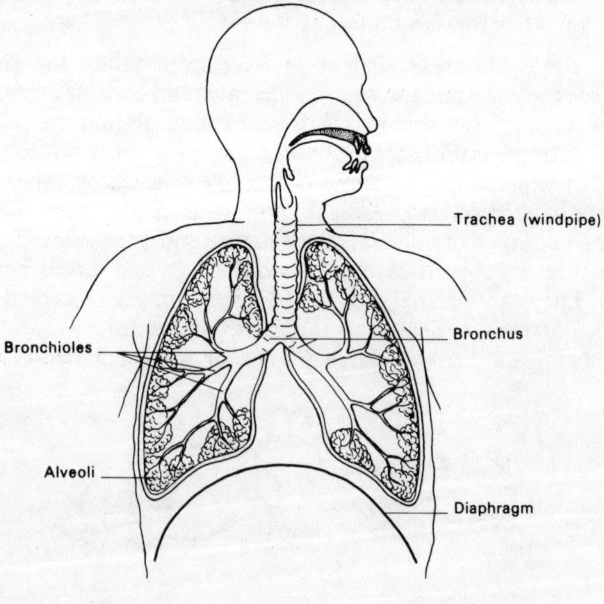

Fig.12.3 The lungs

The oxygenated blood returns through the *pulmonary vein* to the heart where it is pumped into the *aorta* and thence into the main arteries and blood vessels of the body. The blood carries oxygen from the lungs and nutrients from the organs of the digestive system to all the cells and tissues of the body. It also carries away the soluble waste products of tissue activity which are disposed of via the lungs as carbon dioxide, through the kidneys as urine or through glands in the skin as sweat. Oxygenated blood is bright red and is carried through the *arteries* of the body by the pumping action of the heart. The position of an artery may be found by feeling for this beating or pulse. Deoxygenated blood is dark red or blueish-purple and is returned from the tissue capillaries to the heart in the *veins*, which (unlike the arteries) do not beat. The normal adult pulse rate at rest is about seventy beats per minute.

Respiration and circulation (i.e. breathing and the beating of the heart) are both essential to life. The maintenance of these two systems is thus an urgent and vital priority in first-aid.

(b) *Asphyxia*

A person will suffer permanent brain damage and will probably die if his brain is deprived of oxygen for more than about four minutes. Immediate action is therefore essential if the accident victim has stopped breathing or if—because of the effects of cyanide or carbon monoxide poisoning, for example, or a deficiency of oxygen in the air he is breathing—his blood or tissues are not receiving an adequate oxygen supply. This condition is known as *asphyxia* and is one of the commonest causes of unconsciousness (see 12.17).

A casualty suffering from asphyxia will show some or all of the following symptoms:

1. Difficulty in breathing. The rate and depth of breathing increase at first and may become noisy with frothing at the mouth. The patient may then stop breathing altogether. Artificial respiration (see 12.5) should be applied *immediately* if the victim is not breathing. Any delay could be fatal.
2. Gradual loss of consciousness.
3. Congestion of the head and neck. The face, lips and nail beds of the fingers and toes may become blue. This condition is known as *cyanosis*.

If the victim is showing symptoms of asphyxia it is important to identify the cause quickly. In most cases this will be obvious and once it is removed recovery will be rapid. Asphyxia is an indication that *either* the lungs or heart have ceased to function correctly *or* there is not enough oxygen in the air the victim is breathing. The principal causes of asphyxia are:

1. Obstruction of the airway by a foreign body, blood, vomit, or—in the case of an unconscious casualty lying on his back—by the tongue

falling to the back of the throat and blocking the passage to the lungs. Occasionally this obstruction is caused by the swelling of the tissues of the throat by stings or the effect of swallowed corrosive or scalding liquids.
2. Spasms caused by drowning or the action of smoke or irritating gases.
3. Suffocation by pillows, plastic bags etc. or by compression of the neck or chest.
4. Carbon monoxide or cyanide poisoning. Carbon monoxide is present in the exhaust gases of petrol or diesel engines; these therefore should never be operated in an enclosed space. Carbon monoxide is formed when coke, wood, oil or any carbonaceous fuel is burnt in a limited amount of oxygen. Fires quickly deplete the oxygen supply in a room; this combined with the presence of smoke and toxic or irritant vapours is the major cause of fire deaths (see 5.4).
5. Electric shock (see 1.8 and 12.15). Electrocution can affect the nerves which control respiration.

(c) *Artificial respiration*
The purposes of artificial respiration are:

1. To clear the victim's airway to allow a free passage of oxygen to the lungs
2. To inflate the lungs and ensure the immediate and continual oxygenation of the blood
3. To restart the heart (if it has stopped) in order to maintain sufficient circulation to ensure that oxygenated blood reaches the brain, heart, kidneys and other organs of the body.

12.5 Artificial respiration

Before describing the recommended method of *artificial* respiration, it is worthwhile to consider the process by which *normal* respiration occurs.

(a) *Mechanism of respiration*
Normal respiration is brought about by the action of the *diaphragm* and the muscles between the ribs. During *inspiration* ('breathing in') air is drawn into the lungs by the contraction of the diaphragm and the pulling of the ribs upwards by the action of the muscles attached to them. These two actions increase the capacity of the chest and air is drawn in. In *expiration* ('breathing out'), the air is forced out through the bronchi and trachea by the elastic contraction of the lungs as the diaphragm relaxes and the ribs return to their normal position (fig. 12.4). There is then a short *pause* before inspiration recommences. These movements are controlled by the respiratory centres in the mid-

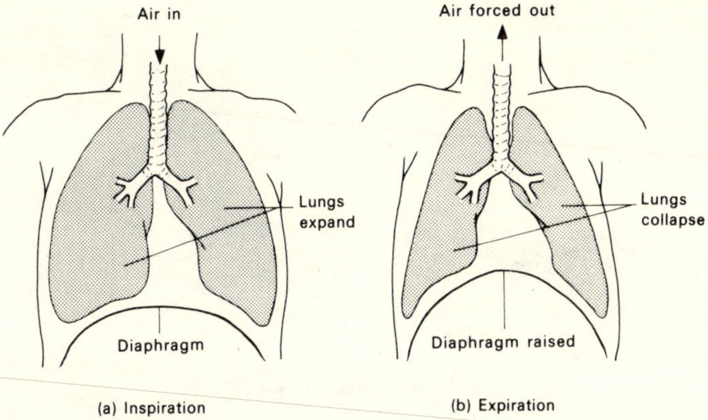

Fig.12.4 Breathing

brain which react to changes in the concentration of carbon dioxide in the blood.

The rate of breathing varies considerably. For an average adult at rest the rate is about 15 to 18 times per minute, while in infants and young children the respiration rate ranges between 25 and 40 times per minute. These normal rest rates increase rapidly during exercise, fever or other conditions (such as pneumonia) which affect normal lung function and increase the demand for oxygen.

(b) *Mouth-to-mouth artificial respiration* (*expired air or 'kiss of life' method*)

If the accident victim has stopped breathing artificial respiration must be started immediately to get a supply of air into the lungs and oxygenate the blood. The most important single factor is the speed with which the first few inflations can be given. Delay can be fatal. The mouth-to-mouth method is the most effective and with few exceptions, such as cyanide poisoning, it can be used in virtually all circumstances. The procedure is as follows:

1. Quickly remove any obvious obstructions covering the head and face or any constrictions round the neck.
2. Ensure that the airway is free. Clear the mouth of any debris, blood, vomit and loose or false teeth.
3. Tilt the head backwards by pressing the top of the head, while supporting the nape of the neck, and push the chin upwards (see fig. 12.5(a)). This extension of the head and neck lifts the tongue forward clear of the airway and in some cases is sufficient to restart the casualty's breathing. If this occurs place him (or her) in the recovery position (fig. 12.2). If the tongue does not fall forward pull it forward with the finger.

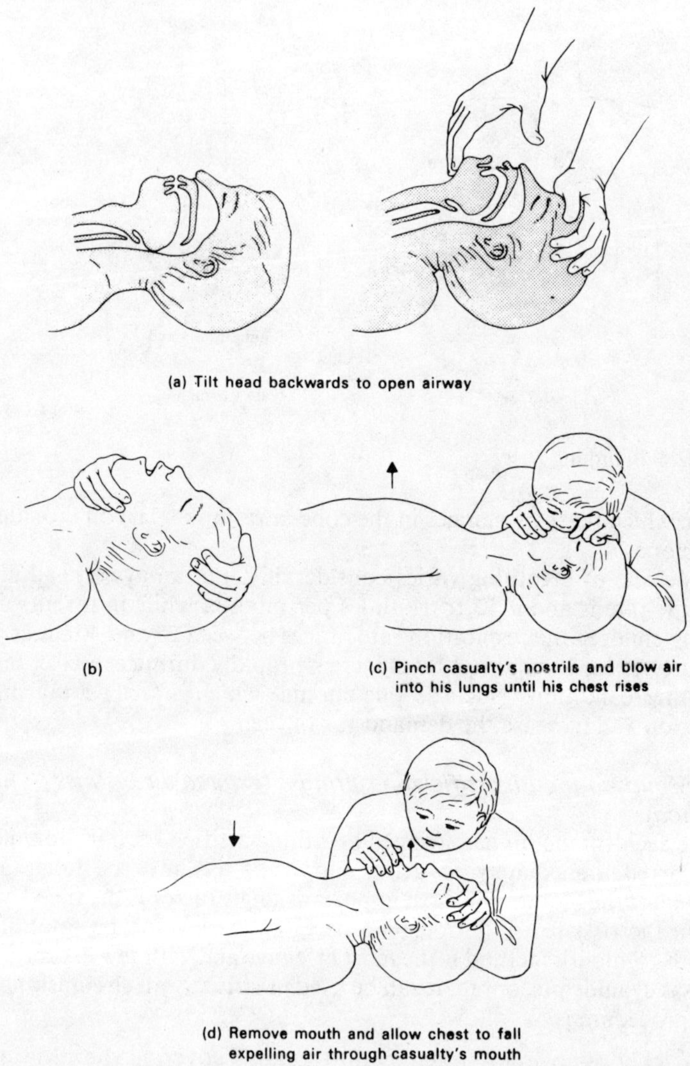

Fig.12.5 Artificial respiration

4. Loosen clothing at the neck and waist.
5. Take a deep breath and open your mouth wide. Pinch the casualty's nostrils together with your fingers and, pressing your lips round his mouth, blow air into his lungs until his chest rises. Remove your mouth and watch the chest fall.
6. Repeat and continue at your natural breathing rate until normal breathing is restored.

Fig.12.6 Dilated pupils

(c) *Heart massage*

In cases of breathing failure the person administering first-aid should check that the casulty's heart is still beating. This is especially important with victims of electric shock or poisoning where heart stoppage is a particular hazard. This check can be carried out by feeling the pulse at the wrist or neck or by applying an ear to the victim's chest. Other symptoms are widely dilated pupils (fig. 12.6) and a grey colour of the skin.

If a heart beat cannot be detected, place the casualty on his back on the floor and strike his chest smartly to the left of the lower part of the breastbone with your fist. If the heart still does not beat, place the heel of the hand at the base of the sternum (or breastbone) keeping the palm and fingers off the chest. Cover the hand with the heel of the other hand as shown in fig. 12.7 and, with arms straight, rock forwards and back-

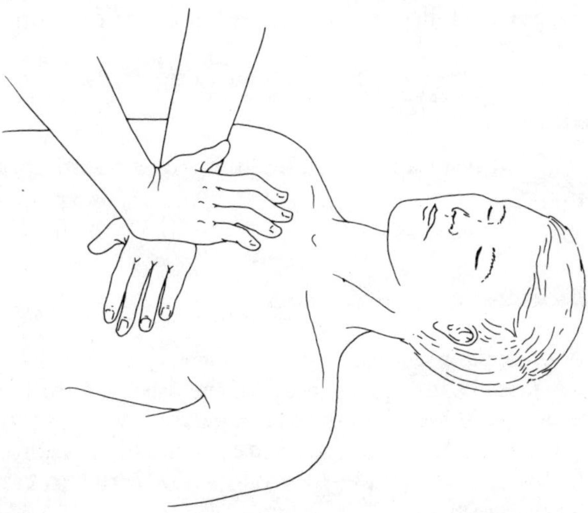

Fig.12.7 Heart massage

Fig.12.8 Dummy for artificial respiration

wards pressing down about 4 cm (1½ inches) at about sixty times a minute. This should be continued *at the same time as artificial respiration* until the heartbeat resumes or until medical help arrives or the casualty reaches hospital. Cycles of fifteen heart compressions followed by two quick lung inflations are recommended if the person administering first-aid is alone. The effective use of this combined technique requires skilled instruction and practice on a dummy (fig. 12.8).

Assignment

Practice artificial respiration with a dummy or a friend until you are proficient.

12.6
First-aid dressings for surface wounds

(a) *Classes of surface wounds*

Any abnormal break in the continuity of the tissues of the body which permits the escape of blood is known as a *wound*. Wounds can be either external or internal. Both categories are potentially dangerous as—in addition to the loss of blood—they may allow germs to enter which could cause infection.

The commonest wounds occurring in laboratory accidents may be divided into the four types listed in table 12.2.

Table 12.2 *Classification of wounds commonly encountered in laboratory accidents*

Class of wound	Cause/description
Clean cuts or incisions	Cuts from broken glass, knives, scalpels, razor blades etc.—can cause heavy blood loss.
Stabs or punctures	Caused by sharp-pointed instruments, such as stilettos, needles, broken glass rods or thermometers. Small entry hole, but the wound can be deep and produce serious internal injuries.
Lacerations	The skin and edges of the wound are torn and irregular. Lacerations are caused by moving machinery, falls, scratches from animal claws, barbed wire etc. Blood loss is generally small, but as dirt is likely to be present there is a risk of infection.
Bruises or contusions	A *bruise* or *contusion* is the name given to bleeding beneath the surface of the unbroken skin. It is thus a minor form of internal bleeding, but it can be severe if a large area is involved. Bruises are generally caused by falls or a blow on the surface of the body. They may not be apparent at first, except perhaps as a red mark. Swelling and the characteristic blue–black colour of a bruise may take some time to develop.

(b) *First-aid dressings*

After cleaning or other suitable treatment a wound should be covered with a protective dressing, the main purposes of which are:

1. to prevent infection,
2. to control bleeding,
3. to absorb any discharge,
4. to reduce further injury.

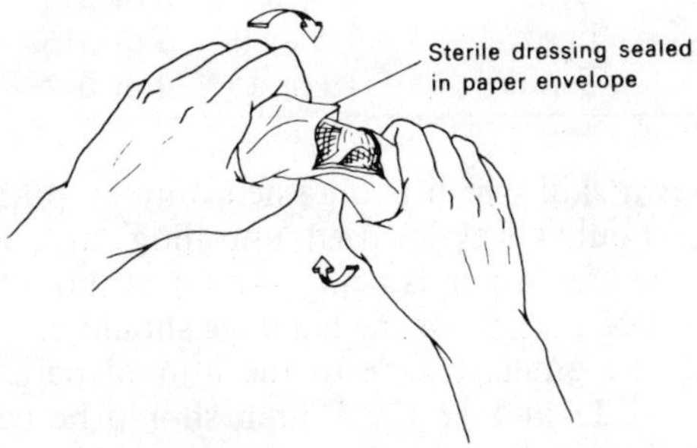

Fig.12.9 Opening a standard gauze dressing

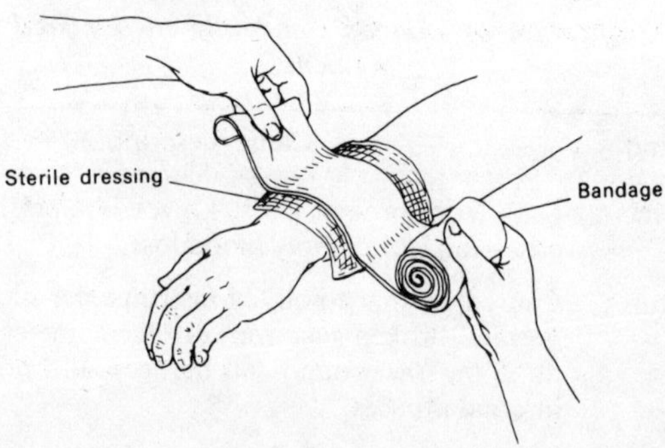

Fig.12.10 Keeping a sterile dressing in place by bandaging

The main types of first-aid dressing for surface wounds are sterile adhesive pads, such as 'Elastoplast' or 'Band-Aid', and the prepared standard gauze dressings covered with a pad of cotton wool which are supplied sealed in paper or plastic covers to keep them sterile (fig. 12.9). Both these types are available in a range of sizes. The adhesive dressings are particularly convenient for small surface wounds. The sterile pad must not be touched with the fingers and the surrounding skin should be dry before applying the dressing. Standard gauge dressing is kept in place by bandaging (fig. 12.10). The width of the roller bandage employed is determined by the part of the body involved:

Part of body	Width of bandage	
	/cm	/in
Finger	2.5	1
Hand	5	2
Arm	5 or 6	2 or 2½
Leg	7.5 or 9	3 or 3½
Trunk	10 or 15	4 or 6

Bandaging is a skill which is obtained only by practice. A tightly rolled bandage should always be used, unrolling only a few centimetres at a time and the part which is being bandaged should be supported. The partly unrolled portion of the bandage should be kept uppermost while applying the outer surface to the injured part. Examples are shown in figs. 12.11 and 12.12. A limb should be bandaged in the position in which it is to remain. Bandages should be applied firmly and should not be too tight or too loose. The tips of the fingers or toes must

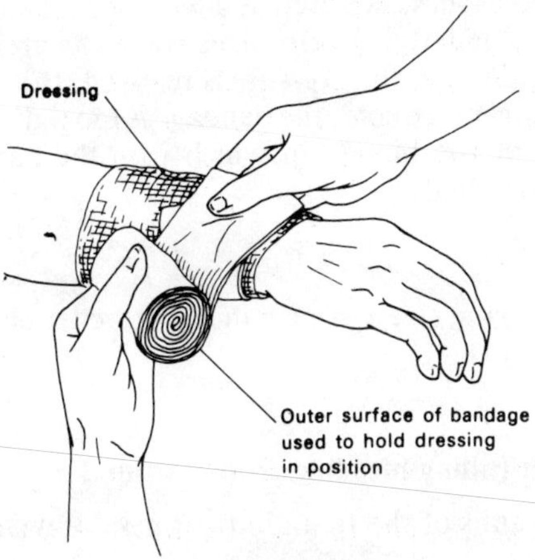

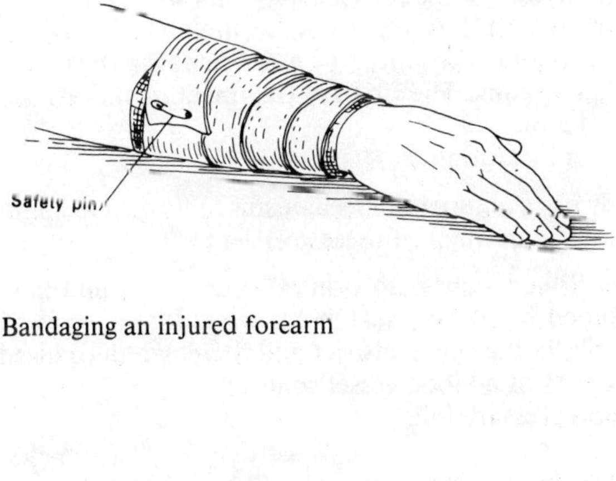

Fig.12.11 Bandaging an injured forearm

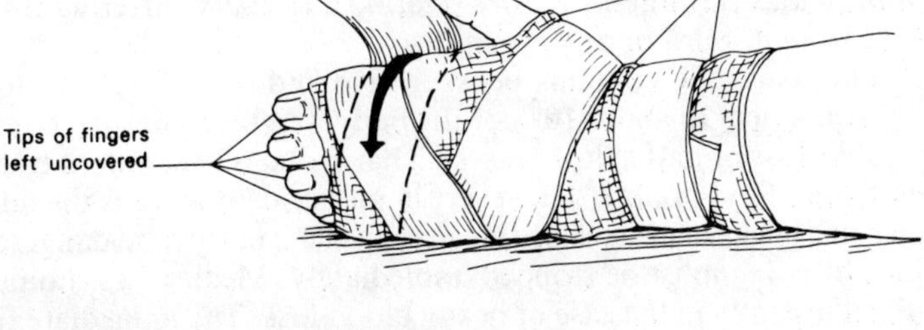

Fig.12.12 Bandaging an unjured hand

never be covered as these are used as a test for circulation. If the nail is pressed until the underlying skin turns white the area should quickly become pink again once the pressure is released. If it remains white or blue, or if the fingers are cold, the bandage is too tight and is restricting the circulation of the blood. In which case the bandage should be removed and reapplied.

Assignment

1 Practise bandaging an arm or a hand. 2 Practise placing an arm in a sling (see fig. 12.21).

12.7
Methods of controlling bleeding from a wound

The immediate aims of the first-aid treatment of wounds are:

1. To control bleeding and to prevent heavy blood loss. Remember, heavy bleeding can kill in less than five minutes—speed is therefore essential.
2. To treat for wound shock. This may vary in severity from a feeling of faintness to death. It results from the diminished activity of the vital body functions and is produced by a deficiency in the blood supply to the important organs. The first-aid treatment of shock is described in sub-section 12.16.
3. To prevent infection.

The body has a number of mechanisms of its own to minimise blood loss. The most important of these are:

1. Clotting. Blood slowly clots on exposure to air and this can block damaged blood vessels and seal minor wounds.
2. The vessels in the skin constrict and thereby reduce bleeding.
3. The cut ends of a blood vessel contract.
4. The blood pressure falls.

Treatment of surface bleeding

Note: Except in cases of emergency where speed is essential (e.g. splashes of acid in the eye), the person administering first-aid should wash his hands thoroughly before treating a casualty suffering from a surface wound, burn or any eye injury.

The human body contains between five and eight litres of blood. This is equivalent to about 10% of the body weight. In normal circumstances the loss of half a litre causes no harm and is the amount usually taken from blood donors. However, in an accident there is the added danger of shock and losses of this magnitude can then be dangerous. Serious bleeding must be stopped immediately. Medical aid should be obtained urgently in the case of heavy blood loss. The immediate first-aid treatment is as follows:

(a) *For heavy blood loss*

1. Control the bleeding by the following means:
   (i) Apply direct pressure on the wound for 5 to 15 minutes with a clean pad of cloth or if this is not available with the fingers or bare hands. Press the sides of large wounds gently but firmly together. Do not apply a tourniquet. If it is not possible to apply pressure directly on the wound, apply indirect pressure at an appropriate point on any artery between the heart and the wound. This treatment prevents blood reaching the wound and is the method which must be applied immediately in any accident in which an artery has been severed. The location of the main pressure points and their use are described in sub-sections 12.8 and 12.9. Indirect pressure may also be applied around the wound using a ring bandage (see fig. 12.13), for example. The technique may be used if jagged pieces of glass or metal are in the flesh.
   (ii) Wherever possible, lay the victim down with the head lower than the rest of the body and—provided an underlying fracture is not suspected—raise the injured part and support it in position. This has the effect of increasing the blood supply to the brain. If the injured part is raised above the heart the pressure effect of having to flow uphill will also reduce blood loss from the wound.
2. Carefully remove any foreign bodies (e.g. broken glass) which can *easily* be picked out of the wound.
3. Apply a dressing directly over the wound and press it down firmly.

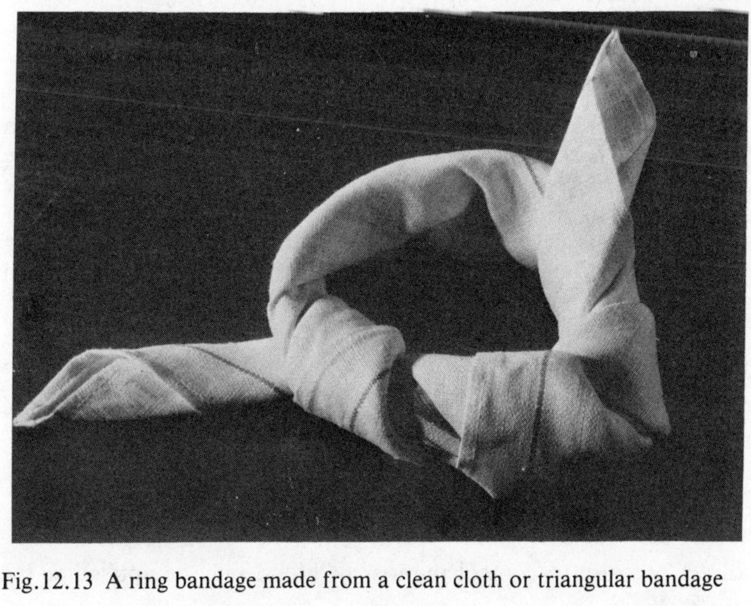

Fig.12.13 A ring bandage made from a clean cloth or triangular bandage

Cover it with a pad of soft material and bind it with a firm bandage to keep the dressing and pad in position.
4. Immobilise the injured part using a sling or, in the case of a lower limb, by padding it and tying it to the other leg.
5. Carefully transport the casualty to hospital.

(b) *For wounds with slight bleeding*
Frequently the bleeding stops of its own accord or is easily controlled by local pressure. The procedure for first-aid treatment is simply as follows:

1. Reassure the casualty and keep him still;
2. Wash the wound in running water. Dry the skin with swabs of cotton wool, using each swab only once and wiping away from the wound;
3. Apply a dressing with a pad if required and bandage firmly. Frequently an adhesive dressing is more convenient;
4. Provided a broken bone is not suspected, raise the injured part and support it in this position with a sling or by resting it on something of a convenient height such as a pillow, table or laboratory stool.

(c) *Internal bleeding*
A serious injury may also produce internal bleeding. The presence of internal bleeding is indicated if the patient spits up or excretes blood, or if there is any swelling, often dark in colour, close to the injury. Blood often appears from the ears or nose in a serious head injury. Medical aid should be obtained urgently if an internal injury is suspected and the casualty should be treated for shock (see 12.16).

12.8
Location of the main pressure points

A *pressure point* is the name given to the places in the body where an artery can be pressed against an underlying bone to control or stop the flow of blood to tissues beyond that point from the heart. Pressure points are used in situations where because of the injury or the nature of the wound it is not possible to control bleeding by the application of direct pressure. The positions of the main pressure points are shown in fig. 12.14.

Assignment

Locate the position of the pressure points in your arm, wrist and leg (see fig. 12.14) by feeling for the pulse at the appropriate point.

12.9
Use of a pressure point

Indirect pressure to control the loss of blood may be applied at the

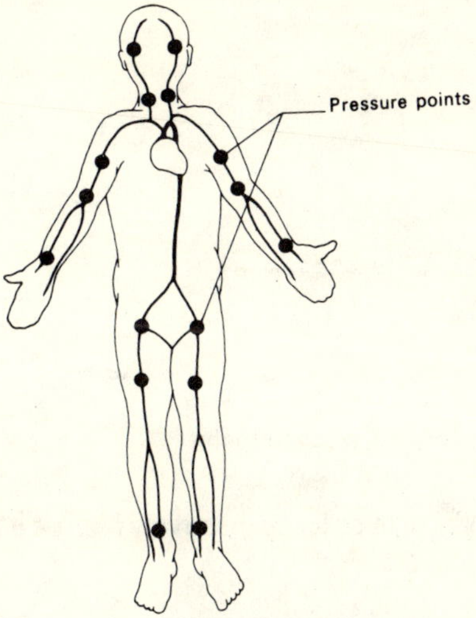

Fig.12.14 Position of the main pressure points

points with the fingers or thumbs. The choice of pressure point is determined by the location of the wound. Pressure should be applied at the pressure point closest to the wound on the path of the artery from the heart. Do not apply pressure on points *beyond* the wound from the heart as this will increase blood loss. Pressure should not be maintained for periods longer than fifteen minutes, but this is usually sufficient while dressings are being prepared for the wound or while waiting for the arrival of an ambulance or doctor. The precise location of the pressure point may be found by feeling for the pulse at the appropriate point. The use of the brachial and femoral pressure points to control the

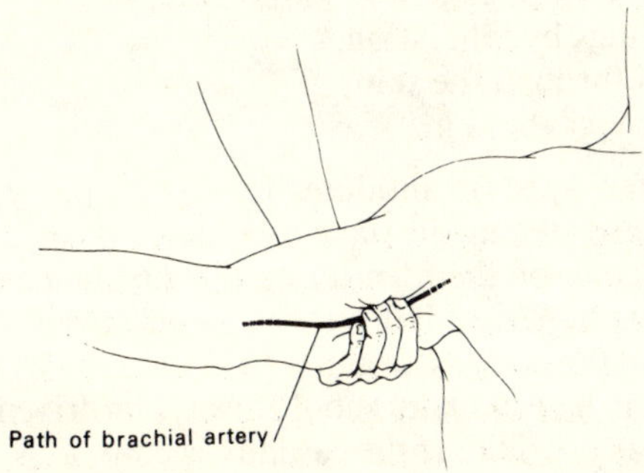

Fig.12.15 Use of the brachial pressure point

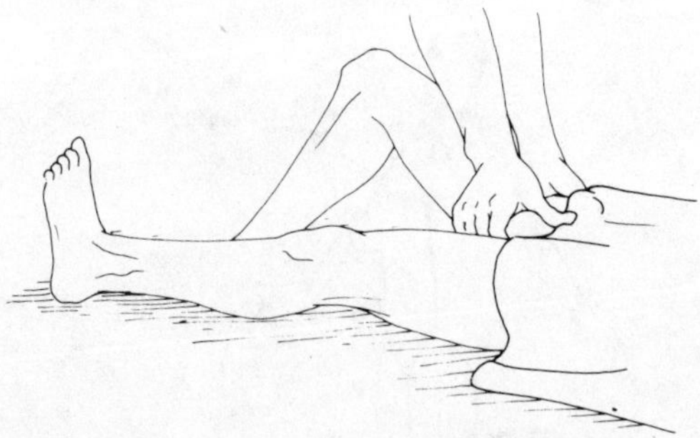

Fig.12.16 Use of the femoral pressure point

flow of blood into the arm or leg respectively is shown in figs. 12.15 and 12.16.

12.10
Procedures for general treatment of poisons

Poisonous substances are a particular hazard of chemical and other laboratories. Most laboratories store hundreds or even thousands of compounds which to a greater or lesser extent are toxic and many more may be produced by these substances in chemical reactions. Provided the basic precautions are taken these materials do not present any great danger, nevertheless anyone working in a laboratory should be familiar with the first-aid treatment to apply if an accident occurs. British Drug Houses (BDH) and a number of other firms publish first-aid charts which describe the immediate treatment to apply in accidents for a large number of different chemicals.

Poisonous substances may be gases, liquids or solids. They can be taken into the body by the following routes:

1. through the mouth by ingestion (swallowing),
2. through the lungs by inhalation,
3. by absorption through the skin,
4. by injection under the skin.

As there are few specific antidotes for the majority of poisons, a number of standard treatments have been devised which apply in any accident, irrespective of the identity of the substance involved. The standard treatment to be used is determined solely by the route by which the poison entered the body. The aim is to sustain life by eliminating or diluting the poison. The casualty should always be driven to the nearest hospital as soon as possible. If the casualty is conscious, ask him what the substances were and give this information to the doctor at the hospital.

Standard treatments

(a) *Ingestion of poisonous substances*
This usually presents the least hazard as it is unlikely that significant amounts of a toxic liquid or solid will be swallowed deliberately. The possibility of this type of accident occurring is reduced considerably if liquids are never pipetted by mouth (see 9.8).

The standard treatment for this type of poisoning is as follows:

1. Tell the victim to spit out as much of the material as possible and then wash the mouth out thoroughly a number of times with water. Half a dozen separate washes with fresh water each time are more effective than a single wash with a large amount of water. Do not let the victim swallow the mouth washings.
2. If the substance has been swallowed give large drinks of water or milk to dilute the chemical in the stomach.
3. Do not induce vomiting as this may result in further damage to the delicate tissues of the upper food passages if the substance is corrosive.
4. Transport the casualty to hospital. Wherever possible the following information should accompany him: (i) the identity of the poison, (ii) the approximate amount and concentration of the chemical consumed, (iii) brief details of the treatment already given.

Experiments with cyanides and other highly toxic substances should never be carried out without having sufficient amounts of the specific antidote immediately available. With substances as poisonous as sodium cyanide or potassium cyanide there will not be time to prepare this solution after an accident has occurred. The specific antidote for cyanide poisoning is:

Solution A	39.5 g of hydrated iron (II) sulphate and 0.8 g of BP citric acid dissolved in 250 cm^3 of water.
Solution B	15 g of sodium carbonate dissolved in 250 cm^3 of water.

Solution A deteriorates slowly on standing and should be inspected from time to time and replaced as necessary.

The immediate treatment if cyanide has been swallowed is to drink about 150 cm^3 of a mixture of equal amounts of Solution A and B. The antidote acts by inducing vomiting and by forming insoluble non-toxic iron complexes with the cyanide ions. A doctor should then be summoned or the victim taken to hospital at once.

(b) *Inhalation of poisonous gases or vapour*
This route of toxic substances into the body is the most dangerous and has the most immediate action. Most poisonous gases, such as chlorine, hydrogen sulphide, ammonia and hydrogen cyanide, are detectable by their odour or by their irritating effect on the tissues of the respiratory

tract. This initial warning should not be ignored as the nose quickly becomes insensitive to smell. For example, hydrogen sulphide is almost as toxic as hydrogen cyanide, but because of the paralysing effect of the gas it appears to have no odour in high concentration.

The standard procedure in gassing accidents is as follows:

1. Remove the casualty from the danger area, provided this can be done without danger to yourself.
2. Loosen the casualty's clothing and administer oxygen if this is available
3. Apply artificial respiration (see 12.5) if breathing has stopped. Do not use the mouth-to-mouth method if the gas responsible is hydrogen cyanide. It is recommended that a capsule of amyl nitrate should be broken and held under the nose of a victim who has inhaled hydrogen cyanide. Again immediate medical attention must be obtained.
4. Place the casualty in the recovery position (see fig. 12.2 and sub-section 12.17) if he is unconscious.
5. Transport the casualty to hospital if the situation warrants it. Give details of the gas responsible and of the treatment given.

(c) *Splashes of chemicals on the skin*
In addition to the absorption of possibly toxic substances, the accidental splashing of chemicals onto the skin can produce burns as a result of the corrosive nature of the substance involved. It can also cause skin disorders such as dermatitis. The immediate treatment to apply is described in sub-sections 12.12 and 12.13.

(d) *Splashes of chemicals in the eye* (see 12.14)

(e) *Injection of substances under the skin*
Accidents of this type are fortunately rare. They can include bites from snakes or other reptiles and insect bites as well as the accidental injection of drugs or biological material from a hypodermic syringe. Special precautions should be taken in laboratories where the possibility of such accidents exists and antidotes or instructions where other special facilities can be obtained should be provided at the first-aid points.

Assignment

What provision is made for the treatment of cases of accidental poisoning in your laboratory? Is it adequate in view of the nature of the work carried out?

12.11
Procedures for first-aid treatment of thermal burns

Dry burns from bench fires or from picking up hot glassware or metals and scalds from steam, boiling water or other hot liquids are a common hazard in all laboratories. Tissue damage results from direct contact with the source of heat or, in the case of damage to the deeper tissues, by a secondary effect involving heat conduction from the damaged areas. The severity of this secondary effect can be reduced by rapidly cooling a burn.

The damage inflicted by a burn varies from a superficial redness of the skin to extensive blistering and death of the underlying tissues, to charring. There is also considerable danger from shock (see 12.16). This is directly related to the extent of the injury and increases rapidly with loss of fluid (plasma) from the surface of the burn and from the escape of blood or plasma into the surrounding tissues where it causes swelling.

The aims of the first-aid treatment of thermal burns are to reduce the local effects of heat, to relieve pain, to prevent infection of the affected area, to replace fluid loss and thereby reduce shock, and to remove a severely injured casualty to hospital as quickly as possible.

The procedure is as follows:

1. Cool the injury as rapidly as possible and alleviate pain by immersing the area in cold water or holding under a running tap. Alternatively the injury may be cooled by applying icepacks.
2. Remove rings, bracelets, boots or anything else of a constrictive nature before swelling occurs.
3. Cover the wound with a dry, sterile dressing.
4. Give small, cold drinks at frequent intervals to a badly burned **conscious** casualty to counteract the effect of fluid loss.
5. Reassure the patient.
6. Badly burned or scalded casualties must be taken to hospital as quickly as possible. Any injury in which more than 10% of the body surface is burned is regarded as very severe and immediate hospital treatment is vital. The figures in table 12.3 provide a convenient guide for quickly assessing the extent of a burn.

Table 12.3 *Guide for estimating relative surface areas of a burn*

Part of the body	Fraction (and approximate percentage) of body surface	
Head	$\frac{1}{9}$	(11%)
Face	—	(3%)
Neck	—	(1%)
Front of trunk	$\frac{1}{9}$	(11%)
Genitals	—	(1%)
Back of trunk	$\frac{1}{9}$	(11%)
1 arm	$\frac{1}{9}$	(11%)
1 leg	$\frac{2}{9}$	(22%)

Do not prick any blisters which form.

Do not touch the affected area as this can increase the risk of infection.

Except in the case of some very minor burns where a little burn cream, such as 'Burnol' or 'Savlon', is a useful immediate treatment to ease pain, do not apply lotions or ointment etc.

In accidents in which a person's clothing catches fire, the immediate treatment is to extinguish the flames by wrapping him in a blanket or coat (but NOT a nylon labcoat) flat on the floor. Smothering the flames in this way puts out the fire by excluding air. Do not remove this covering unless you are certain that the victim's clothing will not reignite on exposure to air. Alternatively the flames may be quenched with cold water.

12.12
Procedures for first-aid treatment of chemical burns

Chemical burns are caused by the action of corrosive or caustic substances on the skin. Examples of such chemicals include phenol, bromine and the strong acids and alkalis (especially concentrated sulphuric (VI) acid, nitric (V) acid, sodium hydroxide (caustic soda) and potassium hydroxide). The common antidotes for the treatment of burns from these substances are described in sub-section 12.13. The standard first-aid treatment for chemical burns is the same as that for dealing with splashes of poisons or other potentially hazardous chemicals on the skin.

The procedure is as follows:

1. Drench the affected area with large amounts of running water. Continue for at least five minutes or until you are satisfied that none of the chemical remains in contact with the skin. Chemicals known to be insoluble in water can be removed with soap under a running tap. In cases where the water supply is limited it is best to wipe as much as possible of the acid or other corrosive liquid from the skin quickly with a clean cloth before using the little water which is available to wash the affected area.
2. Carefully remove all contaminated clothing.
3. If the casualty is seriously injured or if the burn was caused by splashes of hydrogen fluoride or other extremely dangerous substances arrange immediate transportation to hospital or to the nearest doctor.

12.13
The common antidotes for chemical burns by acids, alkalis, phenol, bromine and phosphorus

The treatment described in sub-section 12.12 is applicable to all chemical burns. However, the effects of burns from acids, alkalis, phenol, bromine or phosphorus are considerably reduced and the accompanying pain of the injury is lessened by applying the relevant antidote to remove or neutralise the substance.

(a) *Acid burns*
Wash the affected area thoroughly with water and then with an aqueous 10% solution of sodium bicarbonate to neutralise the acid.

(b) *Caustic alkali burns*
Wash thoroughly with water and then with an aqueous 1% solution of ethanoic (acetic) acid to neutralise the alkali.

(c) *Phenol*
Phenol (or carbolic acid) is a white crystalline solid, melting point 43 °C, which slowly turns pink on exposure to air. It has a sharp, disinfectant-like odour. Phenol is both toxic and caustic and it is therefore important to prevent the substance being absorbed through the skin into the underlying tissue. The affected area should be washed thoroughly with water and then swabbed with glycerol to remove any phenol absorbed in the skin.

(d) *Bromine*
Bromine is a reddish-brown fuming liquid, boiling point 59 °C. It gives off a very poisonous vapour which irritates the eyes and mucous membranes and causes severe burns if it comes into contact with the skin. Burns caused by splashes of bromine should be washed thoroughly with water and then neutralised with dilute aqueous ammonia, sodium bicarbonate or aqueous 5% sodium thiosulphate solution. Bromine can also be removed from the skin by washing with a small amount of petrol.

(e) *Phosphorus*
Phosphorus burns may be treated by washing the affected area with an aqueous 3% solution of copper sulphate.

12.14
The use of eye-wash bottles

The eyes are a particularly valuable (and vulnerable) part of the body. Transplants of a complete eye are not possible and the sight of an eye injured in an accident, once lost, can rarely be restored. It is not an exaggeration therefore to say that you are reading this book with your last pair of eyes. Artificial legs or limbs provide some degree of mobility but it is not possible to see with a glass eye. The importance of wearing goggles or a face shield for any experiment or workshop operation where there is any danger of splashes of chemicals, broken glass or

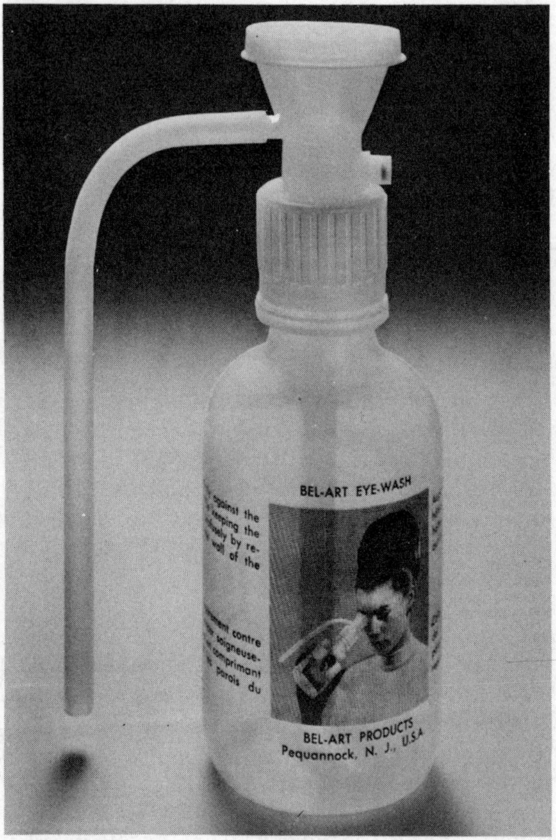

Fig.12.17 An eye-wash bottle

particles of metal entering the eye cannot be stressed too strongly (see 7.1). It is far easier to prevent eye damage than it is to cure it once it has happened.

Splashes of chemicals or of corrosive liquids in the eye must be treated immediately as any delay may result in permanent damage to the sight. Strong alkalis are particularly dangerous. The aim of the first-aid treatment is to dilute and eliminate the chemical as quickly as possible and then to get the casualty to hospital for urgent treatment.

The first-aid procedure is as follows:

1. Hold the eye open or get the victim to blink repeatedly while washing the eye with water from an eyewash bottle (fig. 12.17) or from a gently running tap for several minutes. *Note*: Eye injuries frequently produce a spasm which keeps the lids firmly shut and considerable care and skill are required to gently prise them apart with the fingers to ensure that the water reaches the eyeball.

2. Place a clean dressing over the eye.
3. Arrange immediate transport to hospital. All eye injuries caused by the action of chemicals require urgent medical treatment. Don't send the victim home until this has been done even if he insists that he is all right as in some cases the effects of the injury may not develop for some time.

Foreign bodies, such as a piece of grit or an eyelash, may be removed from the eye using the corner of a *clean* handkerchief. All eye injuries resulting from solid objects should receive urgent skilled medical attention.

Assignment

Examine the eye-wash bottle supplied in the laboratory and ensure that you are able to use it correctly both on yourself and on another person. How does it work? Try it out, but remember to refill the bottle after use with clean water.

12.15
First-aid treatment for electric shock

The physiological consequences of the passage of an electric current through the human body were described in sub-section 1.8. The principal injuries which may be expected in an electrical accident are burns, asphyxia and shock. These may be accompanied by cuts, fractures or other injuries resulting from falls etc. by the sudden muscle spasm produced by the electric current. The sequence of action to be taken is:

1. Do not touch the casualty until you are certain the power has been turned off or that he is no longer in contact with the electric current or you may be electrocuted as well. No attempt at rescue must be made if the victim is in contact with a high voltage electric current such as that coming from overhead electric power cables or some industrial power supplies. With a mains supply voltage (220–240 V) the victim can be pulled or pushed clear using a dry walking stick (not a metal one), a wooden chair, thick dry cloth, rubber or other insulating material. The person administering first-aid should stand on a *dry* insulating surface such as a wooden chair or a pile of books when attempting this.
2. If the victim is not breathing, apply artificial respiration immediately (see 12.5).
3. If the victim is unconscious but breathing place him in the recovery position (see fig. 12.2 and sub-section 12.17).
4. Treat burns (see 12.11) and other injuries as appropriate (see 12.6–12.9 and 12.18).
5. Treat for shock (see 12.16).
6. In cases of serious injury send for an ambulance or doctor.

12.16
First-aid treatment of wound shock

Shock is a state of collapse produced by an insufficient blood supply to the vital organs of the body. It may occur as a result of severe external or internal bleeding, a heart attack, fractures or by the loss of blood plasma from circulation by extensive burns. Some degree of shock is produced in almost every accident and is present even in cases where the victim simply suffers severe fright, but no actual physical injury. In some laboratory accidents, such as explosions or burns from splashes of concentrated sulphuric (VI) acid, the effects of shock may be more serious than the injuries themselves. Shock varies enormously in its severity and can be fatal. Sometimes the effects are delayed and shock may not be apparent until some time after the accident.

The chief symptoms of shock are as follows:

1. Paleness and a cold, clammy skin. This is caused by the contraction of the surface blood vessels as the blood is directed to the internal organs. The victim may also sweat profusely.
2. Shallow and rapid breathing as the victim gasps for breath.
3. An increased pulse rate. Sometimes the pulse also becomes weak.
4. Trembling and faintness. The victim may feel giddy and complain of blurred vision.
5. A feeling of sickness and vomiting.
6. The victim may be very anxious.

The procedure for the first-aid treatment of wound shock if the victim is seriously injured is to get him to the hospital at once. Do not waste time as an immediate blood or plasma transfusion may be required if his life is to be saved.

In other cases the following treatment should be given:

1. Lay the victim down or get him a chair if he is not too seriously injured and deal with the injury or other cause of the shock. Protect against cold or draughts by wrapping him in a blanket, but do *not* use hot water bottles or electric fires as this draws blood from the vital organs to the skin.
2. Loosen ties, belts or other tight clothing at the neck, chest and waist.
3. Raise the victim's lower limbs, if possible, to bring more blood back into the brain (see fig. 12.18). If the victim is sitting down, e.g. after fainting, it may be more convenient to get him to lower his head between his knees.
4. Keep the victim still and calmly assure him that all is well.
5. Do not move the patient unnecessarily.
6. If the victim complains of thirst, moisten his lips with water. Remember, a seriously injured person may require an anaesthetic immediately on arrival at hospital so do not give him anything to drink until you are satisfied that any injury is only trivial, e.g. faintness or

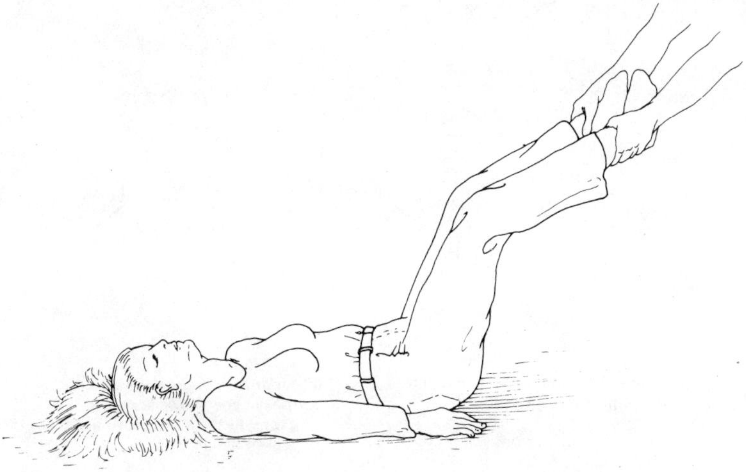

Fig. 12.18 Raising victim's lower limbs

dizziness. Do not allow the victim to drink alcohol and never give tea, coffee or any other liquid to a person who is unconscious or is suspected of having internal injuries.

12.17
Unconsciousness

Unconsciousness (or coma) is a result of some interference with the function of the nervous system and circulation, and is due to an interruption of the normal activity of the brain. The commonest causes of unconsciousness in a laboratory accident are: asphyxia (see 12.4 and 12.5), fainting, shock (see 12.16), poisoning (see 12.10) and injuries to the head. Other causes include heart attacks, epilepsy and strokes.

The general procedure to be adopted for the first-aid treatment of an unconscious casualty is:

1. Remove false teeth and clear the mouth of blood, mucus etc. with a cloth to ensure that the airway is clear.
2. Loosen clothing about the neck, chest and waist.
3. Remove the casualty from a contaminated atmosphere. Open windows and doors to ensure a supply of fresh air to breathe.
4. If breathing fails or stops apply artificial respiration immediately (see 12.5).
5. Control any severe bleeding (see 12.7–12.9).
6. Dress wounds (see 12.6) and attend to fractures (see 12.18) and other injuries.

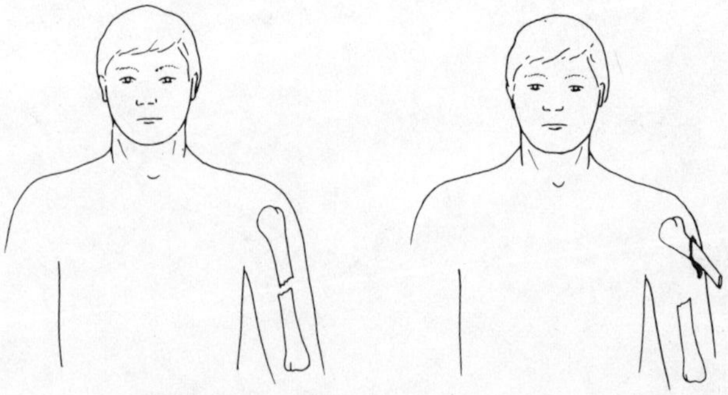

(a) A closed fracture – the skin surface is not broken

(b) An open fracture – the skin is broken and allows access of germs to the soft tissues and broken bones

Fig. 12.19 Fractures of the upper arm

7. Ensure that the casualty's head is kept pressed backward and his lower jaw forward so that the chin juts out as shown in fig. 12.5(a). This prevents the tongue falling back into the throat and choking the victim.
8. Cover the victim with a blanket and arrange for him to be transferred to hospital in the recovery position (see fig. 12.2).
9. Keep a written record of the casualty's responses and pulse rate at regular intervals while waiting for the arrival of the ambulance or of transport to hospital. This information will be useful to the doctor.
10. Keep the casualty still if he regains consciousness. Reassure him and moisten his lips with water but do not give him anything to drink. Advise him to see a doctor.
11. Do not leave an unconscious casualty unattended.

12.18
First-aid treatment for fractures

Any broken or cracked bone is referred to as a *fracture* (fig. 12.19). Occasionally with young children this break is incomplete and is described as a *greenstick fracture* (fig. 12.20). The general symptoms of a fracture are as follows:

1. Tenderness on applying gentle pressure to the affected area and localised pain which increases if the injured part is moved. Some fractures such as those of the wrist or fingers produce little pain and the casualty may feel that he has only bruised or strained the affected area. An X-ray may be needed for confirmation.
2. Swelling. This occurs as a result of blood loss into the surrounding

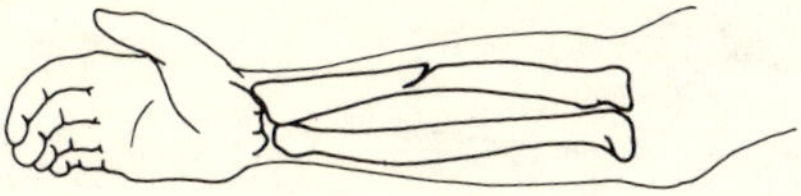

Fig.12.20 A greenstick fracture

tissues. The other symptoms of a fracture may not be apparent because of this swelling, but if in doubt the injury should always be treated as if it were a fracture.

3. Deformity or unnatural movement. Wherever possible the injured and uninjured limbs should be compared. Sometimes the sound of the broken ends of the bone grating against one another may be heard. This is known as *crepitus*.
4. Shock (see 12.16).

The recommended action in any accident in which a fracture is suspected is to keep the casualty still and not to move him unless it is necessary to separate him from some other hazard which could endanger his life. Remember that any movement can cause further injury and the part should be immobilised by means of a body bandage or by the use of splints and bandages. The body bandage uses the casualty's own body as a means of support (fig. 12.21). A broken leg

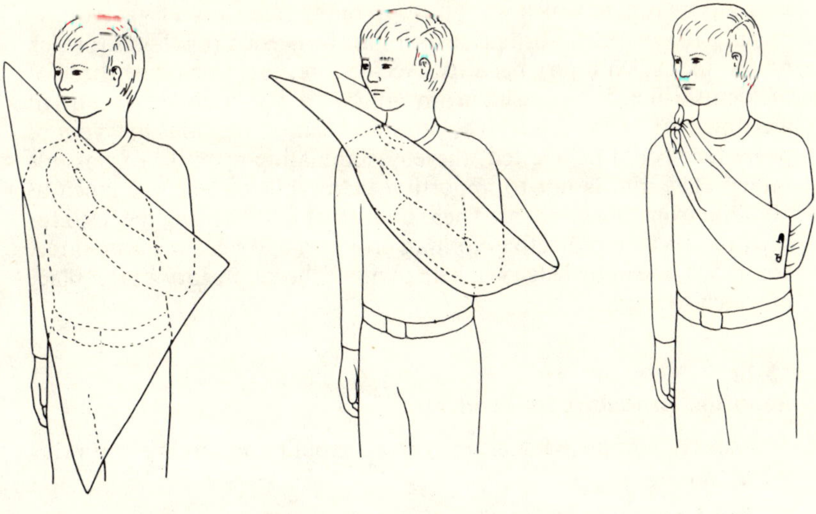

(a) (b) (c)

Fig.12.21 Use of triangular sling to support hand and forearm in case of hand injury or possible fracture of ribs, arm or wrist

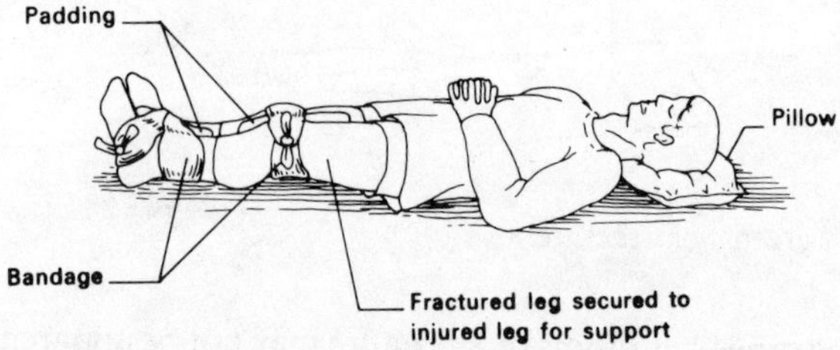

Fig.12.22 Bandaging a fractured leg for support

can be immobilised by first padding it and then bandaging it to the uninjured leg for support. These methods are shown in figs. 12.21 and 12.22. An unskilled first-aider should only attempt these in an emergency when it is impossible to get an ambulance or skilled medical assistance or if it is necessary to move the casualty. The best general treatment by someone lacking this skill is to cover the casualty with a blanket to keep him warm and send for a doctor or an ambulance. Do not give the casualty anything to drink. Treat for shock (see 12.16).

12.19
The need for reporting laboratory accidents

It is important to keep a record of all laboratory accidents and of any first-aid treatment which was given. In many cases it is a legal requirement to record this information as it may be needed for claims for sick pay or industrial injury benefits. Accident reports (including those of incidents which do not result in any injury) are also useful as a means of alerting staff to dangers which with suitable modifications to layout or procedure could be averted, thereby eliminating a source of possible injury. The aim is not to apportion blame but to learn as much as possible from any accident. Then, by further emphasising the need for staff to wear the protective clothing provided and become more safety conscious, the knowledge will help to make the laboratory a safer place in which to work.

12.20
Reporting procedure for accidents

A properly completed accident report should contain the following information:

1. The date, time and location of the accident.
2. The name, address and age of the casualty. The age is important if the victim is under eighteen, as the consent of the parent or guardian is required before a child less than sixteen years of age can be given treat-

ment. However in circumstances such as these, the agreement of the teacher or headmaster, who for the purposes of medical treatment in accidents has the legal authority of a temporary guardian, is sufficient.

3. A description of the accident.
4. The names of any witnesses.
5. Details of any injuries suffered or suspected.
6. A description of the first-aid treatment given.
7. The name of the person who gave the treatment.
8. A description of any further action taken (e.g. removal to hospital).

The use of a standardised report form for all accidents is recommended. Once the report is completed it is torn out of the book and forwarded to the relevant authority leaving the carbon copy as a permanent accident record.

Assignment

What is the procedure for reporting accidents in (a) the laboratory where you work and (b) the college laboratory? Can you suggest any improvements?

Questions: *First-aid*

12.1 What is meant by the term 'first-aid'?
12.2 What are the aims of first-aid?
12.3 Why is it essential for a technician to have some knowledge of first-aid?
12.4 Why aren't first-aid boxes fitted with locks?
12.5 What determines the most suitable site for a first-aid box?
12.6 What are the essential contents of a first-aid box for laboratory use?
12.7 State the priorities in administering first-aid.
12.8 What are the functions of the lungs and heart?
12.9 What are the purposes of artificial respiration?
12.10 What is asphyxia?
12.11 The blood emerging from a cut vessel is bright red and appears in 'pulses'. Is the cut vessel a vein or an artery?
12.12 Why do a person's pulse rate and breathing rate increase on exercising?
12.13 What is a wound?
12.14 What are the main reasons for putting a dressing on a surface wound?
12.15 What are the methods used for controlling bleeding from a surface wound?
12.16 Give the locations of *three* pressure points in the human body. What are their functions?
12.17 What are the *four* routes by which poisons may enter the body?

12.18 What are the standard procedures for the general treatment of poisons?
12.19 What is the procedure for the first-aid treatment of thermal burns?
12.20 Why is it necessary to remove rings as quickly as possible from a burned finger?
12.21 What is the procedure for the first-aid treatment of chemical burns?
12.22 What action would you take if the person working beside you in the laboratory received splashes of chemicals in both his eyes?
12.23 What is (a) the cause and (b) the treatment of wound shock?
12.24 What is the *first* action to take on finding a person in contact with a live electric cable?
12.25 What is the first-aid procedure to be adopted when treating an unconscious casualty?
12.26 Why is it important to immobilise a casualty who has suffered a possible fracture?
12.27 Why is it important to place an unconscious casualty in the recovery position rather than leaving him flat on his back?
12.28 Why is it important not to give a seriously injured person cups of tea or alcohol?
12.29 Why is it necessary to keep written records of any accidents?
12.30 Describe a suitable procedure for reporting accidents which occur in a laboratory.

Further reading

G.D. Muir (ed.), *Hazards in the Chemical Laboratory*, The Chemical Society, London, 2nd edn, 1977.

W. Handley (ed.), *Industrial Safety Handbook*, McGraw-Hill, Maidenhead, 1977.

Safety in science laboratories, Department of Education and Science Series No. 2, HMSO, London, 2nd edn, 1976.

Safety Measures in Chemical Laboratories, National Physical Laboratory, HMSO, London, 1964.

L. Bretherick, *Handbook of Reactive Chemical Hazards*, Butterworths, London.

Code of Practice for the Protection of Persons Exposed to Ionising Radiation in Research and Teaching, Department of Employment, HMSO, London.

Notes for the Guidance of Schools, Establishments of Further Education and Colleges of Education on the Use of Radioactive Substances and Equipment Producing X-rays, DES, HMSO, London, 1976.

The Use of Ionising Radiations in Educational Establishments, DES Administrative Memorandum AM 2/76, HMSO, London.

Radiological Protection in Universities, Committee of Vice-Chancellors and Principals of the Universities of the United Kingdom, 36 Gordon Square, London WC1, 1972.

Guide on the Protection of Personnel Against Hazards from Laser Radiation, BS 4803 : 1972, British Standards Institution, London.

The Use of Lasers in Schools and Other Educational Establishments, DES AM 7/70, HMSO, London. *Laboratory use of Dangerous Pathogens*, DES AM 6/76, HMSO, London.

Industrial Safety and Fire Protection, Bell's Security Handbooks, Bell and Sons, London, 1973.

Code of Practice for Chemical Laboratories, Royal Institute of Chemistry, London, 1976.

N.V. Steare (ed.), *Handbook of Laboratory Safety*, Chemical Rubber Company, 1971.

P.J. Gaston, *The Care, Handling and Disposal of Dangerous Chemicals*, Northern Publishers, Edinburgh.

How to Deal with Spillages of Hazardous Chemicals, BDH Wallchart, British Drug Houses.

Laboratory Waste Disposal Manual, Manufacturing Chemists Association.

A.J.D. Cooke, '*A Guide to Laboratory Law*', Butterworths, London, 1976.

Health and Safety at Work, etc. Act 1974, HMSO, London, 1974.

A. Broadhurst, *The Health and Safety at Work Act in Practice*, Heyden, London, 1978.

F. Wrigglesworth and B. Earl, *A Guide to the Health and Safety at Work Act*, The Industrial Society, London, 1974.

The British Standards Yearbook, British Standards Institution, London.

M.L. McGlashan, *Physico-chemical Quantities and Units*, 2nd edn, Royal Institute of Chemistry, London, 1971.

First-Aid, Handbook of St John Ambulance and British Red Cross Society.

Laboratory First-Aid, BDH Wallcharts, British Drug Houses.

D. Hughes, *Hazards of Occupational Exposure to Ultra-violet Radiation*, Occupational Hygiene Monograph No. 1, 1978.

Safety and Health at Work, TUC Handbook.

Objective test

This test is intended for self assessment on the complete unit and should be attempted without reference to the text or to the answers at the back of the book. With 75 questions it is only possible to sample your knowledge and understanding of the unit. 55 correct answers should be regarded as the minimum satisfactory pass standard, but before you can claim a reasonable mastery of the unit content you should list the sections on which the questions were answered incorrectly and re-read the relevant parts of the text. Finally, the general and specific objectives at the beginning of each section of the text may be used as a checklist and ticked individually once the objectives have been achieved.

Section A Multiple choice questions

1 and 2 There are six different ways in which the brown, blue and green/yellow covered wires of a mains lead may be connected to the three pins of a plug top:

	Responses					
	A	B	C	D	E	F
Brown covered wire connected to terminal marked:	E	E	N	N	L	L
Blue covered wire connected to terminal marked:	L	N	L	E	N	E
Yellow/green striped wire connected to terminal marked:	N	L	E	L	E	N
Where E = earth, L = live and N = neutral						

By writing the appropriate letter(s) for the response on your answer sheet, indicate:

1 the correct method of connecting the plug;

2 the two ways listed which could be lethal even before the equipment itself is switched on;

3 The current drawn by a 5 kW heater operating at 250 V is:

A 20 A

B 5250 A

C 125 A

D 50 mA
E 13 A

4 The most appropriate mains plug fuse for an appliance rated at 2000 W is:
A 3 A
B 5 A
C 13 A
D 30 A
E 45 A

5 The following types of radiation are produced by the decay of radioactive isotopes:
A alpha-, beta- and gamma-rays.
B X-rays and laser light.
C ultra-violet and infra-red light.
D gamma radiation and microwaves.

6 The correct order for increasing wavelength of the following types of electromagnetic radiation is:
A microwaves, infra-red, visible light, ultra-violet, X-rays, gamma rays
B gamma-rays, microwaves, X-rays, visible light, ultra-violet, infra-red
C gamma-rays, X-rays, ultra-violet, visible, infra-red, microwaves
D infra-red, gamma-rays, X-rays, visible light, microwaves, ultra-violet

7 The most suitable extinguisher for a fire close to the mains power supply switch in an enclosed room is:
A dry powder
B carbon dioxide
C foam
D carbon tetrachloride
E B.C.F.

8 The most appropriate extinguisher for burning sodium is:
A water
B carbon tetrachloride
C carbon dioxide
D dry sand

9 Which of the following extinguishers should NOT be used on burning oil?
A foam
B carbon dioxide
C carbon tetrachloride
D B.C.F.
E a fire blanket
F water

10 The micrometer reading in fig. OT1 is:
A 1.69 mm

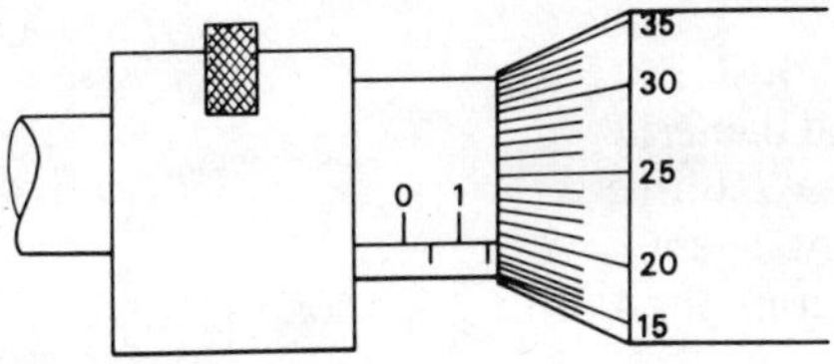

OT1 Micrometer reading for question 10

B 1.71 mm
C 1.21 mm
D 1.19 mm
11. The vernier reading in fig. OT2 is:
A 17.98 cm
B 17.08 cm
C 17.55 cm
D 17.80 cm

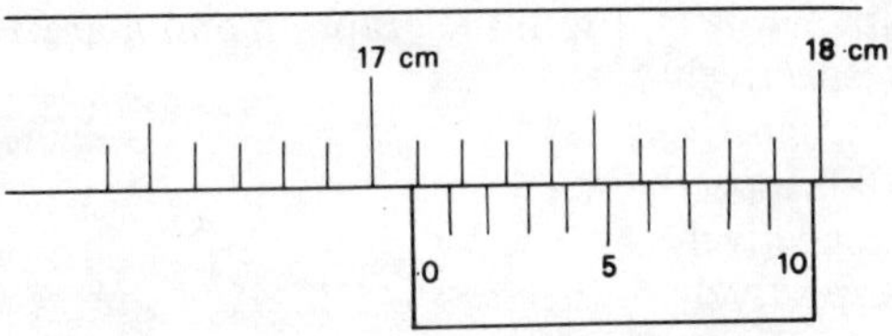

OT2 Vernier reading for question 11

12–16 Pick the appropriate warning symbol from fig OT3 for the following hazards:

OT3 Questions 12-16

12 X-rays
13 an irritant dust
14 pathogenic bacteria
15 a radioactive source
16 an oxidising agent
17 The correct method of lifting a load involves:

	A	B	C	D	E
Keeping the legs straight		✓	✓		
Bending the legs	✓			✓	✓
Keeping the back straight				✓	
Bending the back	✓	✓	✓		✓
Keeping the feet together	✓	✓			
Keeping the feet slightly apart			✓	✓	✓

18 The graph in fig. OT4 was plotted to determine the constants in the relationship: $y = kx^n$. From the graph, n and k are:
A 2 and 2 respectively
B 2 and 100 respectively
C 0.5 and 100 respectively
D 1 and 2 respectively
E 4 and 2 respectively

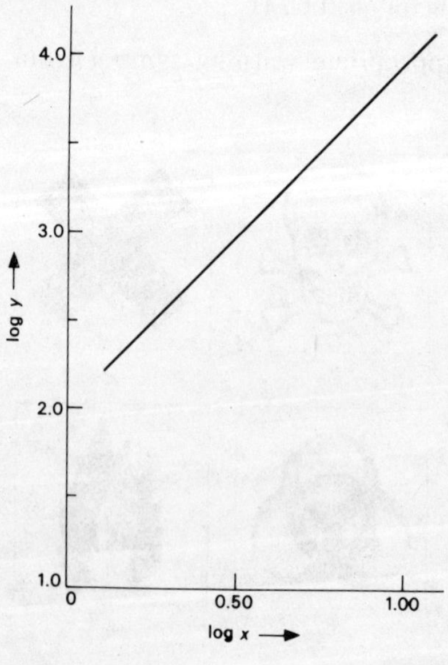

OT4 Question 18

19 A diagram in which the areas of a series of rectangles are proportional to the class frequencies to which they refer is known as:

A A pictogram

B A histogram

C A pictorial bar chart

D A pie chart

E A 100% bar chart

20 To allow for evaporation losses lead–acid batteries and nickel–cadmium cells should be topped up with:

A battery acid

B alkali and sulphuric acid respectively

C distilled water

D sulphuric acid and alkali respectively

E boiled tap water

21 Compared with the lead–acid cell, a nickel–cadmium alkaline cell has the following characteristics:

Reponse	A	B	C	D	E
It is heavier			✓	✓	
It can be left uncharged without harm for long periods	✓	✓		✓	
It can be shorted without harming the cell		✓	✓		✓
It has a lower e.m.f.	✓	✓	✓		✓
The aqueous potassium hydroxide electrolyte is not harmful to the skin or eyes	✓			✓	✓

22 The person working next to you in the laboratory spills some dilute sodium hydroxide on her fingers and then rubs her eye. The immediate first-aid treatment is:

A Phone for the doctor.

B Neutralise the sodium hydroxide in the eye with dilute acid.

C Treat for shock.

D Wash the eye thoroughly with water and then take her to the hospital.

E Place a sterile dressing on the eye and take her to the hospital.

23 The first action to be taken on finding an accident victim in contact with a live electric wire is:

A Take the casualty to hospital.

B Apply artificial respiration.

C Turn off the power supply.

D Treat burns and other injuries.

24 The best place for a first-aid box is:

A Close to the places where accidents might occur.

B In the safety officer's room.

C In an office away from the laboratory so that the contents can be kept clean.

D In the office of a person qualified in first-aid.

25 The best immediate treatment for a thermal burn is:

A Treat for shock.

B Remove blisters.

C Cover the burn with a sterile dressing.

D Cool the affected area in cold water.

26 A person's fingers were blue and cold after his injured arm had been bandaged. The correct action to take is:

A Raise the injured arm above the patient's head.

B Apply pressure at a pressure point in the upper arm.

C Wrap the limb in a blanket and use a hot water bottle to keep the arm warm.

D Remove the bandage.

E Apply a tourniquet.

27 The recommended general sequence for the treatment of the victim of an accident is:

	Response				
	A	B	C	D	E
Treat burns	1	2	3	1	3
Separate the victim from the hazard	2	3	2	3	1
Apply artificial respiration if breathing has ceased	3	1	1	2	2

28 The commonest and most dangerous route by which poisons enter the body is:

A inhalation

B skin absorption

C injection

D swallowing

29 The immediate treatment of a person with a suspected fracture of the leg is:

A Support the victim and help him to walk before his leg becomes stiff.

B Apply direct pressure over the wound in case there is internal bleeding.

C Keep the victim still and send for an ambulance.

D Confirm that the bone is broken by grating the fractured edges together.

E Raise the limb above the head and support it in this position by resting the foot on a stool of a suitable height.

Section B

By writing the appropriate response T or F on your answer sheet indicate whether the statements in questions 30–75 are *true* or *false*.

30 The present BS colour coding for cable and flex makes it possible for a person who is colour blind to safely connect a plug to a flex end.

31 A metal clad appliance may be safely earthed by wrapping a wire attached to its case round the nearest radiator or cold water pipe.

32 The use of multiple adaptors and distribution boards could cause larger currents to be drawn through the cable leading to the socket than would be safe if only a single plug were used.

33 a current of 1 A passing through a man's body from one hand to the other would be perfectly safe provided he was standing on an insulated surface.

34 Double insulated electrical equipment does not have to be earthed.

35 A trailing necktie is perfectly safe when using power tools or operating moving machinery as even if the tie is caught in the machine it will always break before the operator is harmed.

36 Long hair should be tied back and covered before operating any power tool.

37 The damaging effect of ultra-violet light on the eyes is reduced if the source is viewed through a sheet of glass.

38 The rays from radioactive substances can damage the cells which produce red blood corpuscles in the human body and can cause leukaemia.

39 Reflected laser beams are harmless.

40 Low intensity laser light can produce blind spots at the back of the eye as a result of the concentrating effect of the lens in the eye.

41 Lasers are safer in a darkened room.

42 Gamma-rays will pass through thin sheets of aluminium.

43 A strip of photographic film may be used to monitor the dose of radiation received by exposure to a radioactive material.

44 A jet of compressed air is the safest method of removing fragments of broken glass from the spaces between laboratory equipment.

45 Bulk supplies of flammable liquids and cylinders of compressed gases should be stored in a separate building wherever possible.

46 All carcinogenic substances are safe provided they are only used in small quantities of less than 10 g.

47 All mercury surfaces must be covered as the vapour from the liquid metal is harmful.

48 Induction coils generating more than 5 kV are dangerous because they can yield X-rays.

49 Comparatively harmless vapours are occasionally converted into highly toxic substances when the vapour is drawn through the hot zone of a cigarette.

50 Fire extinguishers should be placed close to the exit from a store room or laboratory.

51 Fire doors must be wedged open at all times so that people can easily escape in the event of a fire.

52 The smoke and fumes from a fire cause more deaths than burns from the fire itself.

53 A fuel, heat and a source of oxygen must be present for a fire to occur.

54 Some gases or vapours may be ignited by a spark containing as little energy as 0.0002 J.

55 Always use the lift to escape from a building once the fire alarm sounds as it is quicker than walking down the stairs.

56 Wherever possible, windows and doors should be opened before evacuating a burning building to allow the smoke to escape.

57 The words *flammable* and *inflammable* have the opposite meaning.

58 Strong acids should be neutralised before disposal.

59 Class B burettes are more accurate and more expensive than Class A burettes.

60 An experimental result should be reported to the maximum number of figures obtainable from the electronic calculator used to calculate it.

61 The concave surface of the mirror of an optical microscope is used with artificial light sources (e.g. an electric lamp) and the plane side is used for daylight.

62 Readings of the volume of an aqueous solution should always be taken at the top of the meniscus.

63 An accurate measurement may be found by taking the average of a series of rough readings.

64 The British Standards Institution does not have the power to enforce the adoption of its standards.

65 Primary cells may be recharged when exhausted by passing an electrical current into the cell from an external source.

66 The Health and Safety at Work Act states that the employer may deduct the cost of labcoats, safety spectacles and other protective clothing from a technician's salary.

67 An unconscious casualty should be kept lying on his back until the arrival of the doctor or ambulance.

68 First-aid boxes should be locked to prevent people removing scissors or other equipment which may be required in an emergency.

69 An unconscious casualty must be breathing if his heart is beating.

70 Blood loss from a wound may be controlled by applying pressure at a suitable point on a vein between the wound and the heart.

71 Rings should be removed from burned fingers immediately.

72 A poisonous gas is not harmful as long as you are able to smell it.

73 A casualty with a possible fracture or internal bleeding should not be given anything to drink.

74 An accident victim may die of shock even though his wounds are not particularly severe.

75 Expired air contains more carbon dioxide than oxygen.

Appendixes

Appendix I

SI Units

The seven base units of the International System of units (in French: Systeme International or SI) are:

Quantity	Name	Symbol
length	metre	m
mass	kilogram	kg
time	second	s
electric current	ampere	A
thermodynamic temperature	kelvin	K
amount of substance	mole	mol
luminous intensity	candela	cd

The *metre* is defined as the length which is equal to 1 650 763.73 wavelengths in vacuum of the radiation corresponding to the transition $2p_{10}$ and $5d_5$ of the krypton-86 atom.

The *kilogram* is equal to the mass of the platinum–iridium cylinder (the international prototype kilogram) kept at the Bureau International des Poids et Mesures (BIPM), Sevres, Paris.

The *second* is the duration of exactly 9 192 631 770 periods of the radiation corresponding to the transition between the two hyperfine levels of the ground state of the caesium-133 atom.

The *ampere* is that constant current which, if maintained in two parallel rectilinear conductors of infinite length and negligible cross sectional area placed 1 metre apart in a vacuum, would produce a force between these conductors equal to 2×10^{-7} newton per metre of length.

The unit of thermodynamic temperature (the *kelvin*) is the fraction 1/273.16 exactly of the thermodynamic temperature of the triple point of water.

The unit of the amount of substance (the *mole*) is defined as the amount of the substance which contains as many particles (atoms, electrons, ions, molecules, photons etc.) as there are atoms in exactly 12 grams of pure carbon-12.

1 mole of atoms = relative atomic mass in grams.

1 mole of molecules = relative molecular mass or 'formula weight' in grams.

The unit of luminous intensity (the *candela*) is the luminous intensity in a perpendicular direction of a surface of $\frac{1}{60}$ square centimetre of a black body at the freezing point of platinum under a pressure of 101 325 Pa (1 atm).

The remaining SI units are formed by combinations of these base units. Examples of such derived units are shown in table A1.

Table A1 Derived units

Quantity	Name	Symbol	Units
Activity of a radioactive source	becquerel	Bq	s^{-1}
Electric charge or quantity of electricity	coulomb	C	A s
Absorbed dose of radiation	gray	Gy	J kg^{-1}
Frequency	hertz	Hz	s^{-1}
Energy or work	joule	J	N m
Force	newton	N	m kg s^{-2}
Pressure	pascal	Pa	N m^{-2}
Resistance	ohm	Ω	V A^{-1}
Conductance	siemens	S	A V^{-1} or Ω^{-1}
Potential difference	volt	V	W A^{-1}
Power	watt	W	J s^{-1}

Other examples of derived SI units for commonly used quantities are:

area	m^2
volume	m^3
velocity	m s^{-1}
density	kg m^{-3}
acceleration	m s^{-2}

Multiples and submultiples of a unit are indicated by the addition of the appropriate prefix:

Prefix	Symbol	Factor by which the unit is multiplied	
giga	G	1 000 000 000	$= 10^9$
mega	M	1 000 000	$= 10^6$
kilo	k	1000	$= 10^3$
hecto	h	100	$= 10^2$
deca	da	10	$= 10^1$
		1	$= 10^0 = 1$
deci	d	0.1	$= 10^{-1}$

Prefix	Symbol	Factor by which the unit is multiplied	
centi	c	0.01	$=10^{-2}$
milli	m	0.001	$=10^{-3}$
micro	μ	0.000 001	$=10^{-6}$
nano	n	0.000 000 001	$=10^{-9}$
pico	p	0.000 000 000 001	$=10^{-12}$

For example,

1 mm = 1 millimetre = 0.001 or 10^{-3} m
1 cm = 1 centimetre = 0.01 or 10^{-2} m
1 nm = 1 nanometre = 10^{-9} m
1 μg = 1 microgram = 10^{-6} g.

Appendix II

Important physical constants

General gas constant (R)	$= 8.314$ J K^{-1} mol^{-1}
	$= 0.8205$ l atm K^{-1} mol^{-1}
Speed of light in a vacuum (c)	$= 2.997\ 925 \times 10^{8}$ m s^{-1}
Planck constant (h)	$= 6.626\ 176 \times 10^{-34}$ J s
Avogadro constant (L)	$= 6.022\ 52 \times 10^{23}$ mol^{-1}
Faraday constant (F)	= 96 484.56 C
Molar volume of an ideal gas at s.t.p.	$= 22.4138$ dm^3 mol^{-1}
Standard temperature and pressure (s.t:p.)	= a temperature of O °C (273.15 K) and a pressure of 101 325 N m^{-2} (1 atm = 760 mm of mercury)
Boltzmann constant (k)	$= 1.380\ 54 \times 10^{-23}$ J K^{-1}

Appendix III

Conversion factors

1 atmosphere (atm) = 760 mm of mercury = 760 torr = 101 325 Pa (or N m^{-2})

1 mm of mercury	= 133.32 Pa		
1 calorie (cal)	= 4.1868 J		
ln *x*	= 2.303 $\log_{10}x$		
T/K	= *t*/°C + 273.15		
1 angstrom unit (A°)	= 10^{-8} cm = 10^{-10} m = 0.1 nm		
1 in	= 25.4 mm	1 foot	= 0.3048 m
1 mile	= 1.609 km	1 km	= 0.621 mile
1 acre	= 0.4047 hectare	1 hectare	= $(100\ m)^2$
			= $10^4\ m^2$
			= 2.47 acres
1 pint	= 568.261 cm^3	1 litre	= 1.7598 pints
1 lb	= 453.592 g	1 oz	= 28.35 g
1 kg	= 2.2046 lb	1 ton (UK)	= 1016.05 kg
1 lbf	= 4.448 22 N	1 kgf	= 9.806 65 N
1 lbf in^2	= 6.894 76 kN	1 Btu	= 1.055 06 kJ
		m^{-2}	
1 kW h	= 3.6 MJ	1 horsepower	= 0.7457 kW
1 Therm	= 105.506 MJ	1 erg	= 10^{-7} J

Appendix IV

Relative atomic masses

Element	Symbol	Atomic number	Relative atomic mass
Hydrogen	H	1	1.0080
Helium	He	2	4.0026
Carbon	C	6	12.11
Nitrogen	N	7	14.0067
Oxygen	O	8	15.9994
Fluorine	F	9	18.9984
Sodium	Na	11	22.9898
Magnesium	Mg	12	24.305
Aluminium	Al	13	26.9815
Silicon	Si	14	28.086
Phosphorus	P	15	30.9738
Sulphur	S	16	32.06
Chlorine	Cl	17	35.453
Potassium	K	19	39.102
Calcium	Ca	20	40.08

Element	Symbol	Atomic number	Relative atomic mass
Titanium	Ti	22	47.90
Chromium	Cr	24	51.996
Manganese	Mn	25	54.9380
Iron	Fe	26	55.847
Nickel	Ni	28	58.71
Copper	Cu	29	63.546
Zinc	Zn	30	65.37
Bromine	Br	35	79.904
Silver	Ag	47	107.868
Tin	Sn	50	118.69
Iodine	I	53	126.904
Barium	Ba	56	137.34
Platinum	Pt	78	195.09
Gold	Au	79	196.967
Mercury	Hg	80	200.59
Lead	Pb	82	207.2

Appendix V

IUPAC nomenclature

The International Union of Pure and Applied Chemistry (IUPAC) has recommended the adoption of a systemic nomenclature for all compounds, however many common substances are still known by their trivial names. For example:

Traditional (trivial or current) name	IUPAC — recommended name
Acetaldehyde	Ethanal
Acetamide	Ethanamide
Acetic acid	Ethanoic acid
Acetic anhydride	Ethanoic anhydride
Acetone	Propanone
Acetyl chloride	Ethanoyl chloride
Acetylene	Ethyne
Carbon tetrachloride	Tetrachloromethane
Chloroform	Trichloromethane

Traditional (trivial or current) name	IUPAC — recommended name
Diethyl ether ('ether')	Ethoxyethane
Ethyl alcohol	Ethanol
Ethyl bromide	Bromoethane
Ethyl chloride	Chloroethane
Ethyl iodide	Iodoethane
Ethylene	Ethene
Formaldehyde	Methanal
Formic acid	Methanoic acid
Methyl alcohol	Methanol
Methylene dichloride	Dichloromethane
Oxalic acid	Ethanedioic acid
iso-Propyl alcohol	Propan-2-ol
n-Propyl alcohol	Propan-1-ol
Toluene	Methylbenzene
Urea	Carbamide
Vinyl chloride	Chloroethene

Appendix VI

Hazchem code

The Hazchem code is one of the systems used to indicate the hazards associated with a particular chemical and to give advice about the emergency procedure to be adopted for disposing of a major spillage which might occur, for example, in a road accident involving a tanker or a lorry load of chemicals. The code is marked on the chemical container or on stores or road vehicles carrying the hazardous material. Warning diamonds incorporating the relevant chemical hazard warning symbol (see 6.5) should also be displayed (see fig. A1).

The Hazchem code consists of two or three symbols: the figures indicate whether water, foam or spray etc. may be used to dispose of the material:

	water	foam	spray	dry agents only
Code 1	✓	✓	✓	✓
2	no	✓	✓	✓
3	no	no	✓	✓
4	no	no	no	✓

The letter indicates whether the material may be washed into drains, whether it is explosive and whether full protective clothing or breathing apparatus have to be worn. The symbol 'E' indicates that the immediate area should be evacuated.

	P	R	S	T	W	X	Y	Z
May be washed into drain	✓	✓	✓	✓	no	no	no	no
Full protective clothing required	✓	✓			✓	✓		
Breathing apparatus to be worn			✓	✓			✓	✓
Explosive material	✓	no	✓	no	✓	no	✓	no

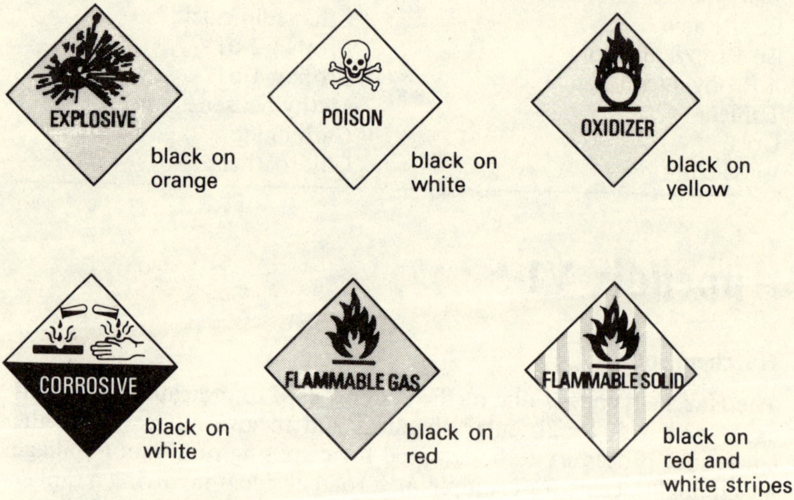

A1 Hazchem warning diamonds

Appendix VII

Conductivity of aqueous potassium chloride solutions

	Temperature /°C	Concentration /mol dm^{-3} 0.001	0.01	0.10
Electrolytic conductivity /S m^{-1}	18	0.0127	0.122	1.12
	25	0.0147	0.141	1.29

Appendix VIII

Primary standards for volumetric analysis

Substance	Formula	Relative molecular mass
Acid–base titration		
Aminosulphonic acid	NH_2SO_3H	97.09
Benzoic acid	$C_6H_5CO_2H$	122.12
Constant boiling point hydrochloric acid*	HCl	36.46
Disodium tetraborate (borax)	$Na_2B_4O_7 10H_2O$	381.37
Sodium carbonate	Na_2CO_3	105.99
Argentimetric titrations		
Silver	Ag	107.87
Silver nitrate	Ag NO_3	169.87
Sodium chloride	Na Cl	58.44
Redox titrations		
Ammonium iron (II) sulphate	$(NH_4)_2SO_4$ $FeSO_4$ $6H_2O$	392.14
Sodium ethanedioate (sodium oxalate)	$Na_2C_2O_4$	134.00
Potassium dichromate (VI)	$K_2Cr_2O_7$	294.20
Potassium iodate (V)	KIO_3	214.01

* The composition of the constant boiling point hydrochloric acid–water mixture depends on the atmospheric pressure:

Pressure, /mm of mercury:	730	740	750	760	770	780
Composition, % HCl by mass:	20.293	20.269	20.245	20.221	20.197	20.173

Appendix IX

Optical density of aqueous potassium chromate (VI) solution

The solution used for the calibration contains 0.0400 g of potassium chromate (VI) in 0.05 mol/l potassium hydroxide.

Wavelength /nm	Optical density	Wavelength /nm	Optical density
215	1.4319	360	0.8296
220	0.4559	370	0.9914
230	0.1675	380	0.9281
240	0.2933	390	0.6840
250	0.4962	400	0.3872
260	0.6344	410	0.1972
270	0.7448	420	0.1261
280	0.7235	430	0.0841
290	0.4295	440	0.0535
300	0.1518	450	0.0325
310	0.0458	460	0.0173
320	0.0620	470	0.0083
330	0.1457	480	0.0035
340	0.3143	490	0.0009
350	0.5528	500	0.0000

Ammonium cobalt (II) sulphate, $(NH_4)_2SO_4\ CoSO_4\ 6H_2O$, or copper (II) sulphate, $CuSO_4\ 5H_2O$, solutions may be used for optical density calibrations of visible region spectrophotometers in the range 350–750 nm.

Appendix X

Density of pure water

The figures have been calculated to allow for buoyancy corrections for weighing in air using brass weights and give the apparent densities and volumes of glassware at 20 °C.

Temperature /°C	Density /g dm^{-3}	Volume of 1 g of water /cm^3
18	997.51	1.0025
19	997.36	1.0026
20	997.19	1.0028
21	997.00	1.0030

Temperature /°C	Density /g dm^{-3}	Volume of 1 g of water /cm^3
22	996.79	1.0032
23	996.59	1.0034
24	996.38	1.0036
25	996.17	1.00385
26	995.93	1.0041
27	995.69	1.0043
28	995.43	1.0046

Appendix XI

Laboratory emergency information sheet

A notice containing the following information should be prominently displayed close to the telephone and first-aid points in a laboratory.

Room or laboratory:
Safety officer:
Deputy safety officer:
Person responsible for maintenance of first-aid box:
Dial 999 in an emergency and state which service (Ambulance / Fire Brigade) you require. Remember to give the exact location of the fire/accident/casualties.

Nearest hospital:
Telephone:

Nearest ambulance service:
Telephone:

Nearest fire service:
Telephone:

Nearest doctors: (i) **Telephone:**
(ii) **Telephone:**

Other services: Laboratories containing appreciable amounts of radioactive substances or other highly dangerous materials should list the names, addresses and telephone numbers of specialist services who should be contacted in an emergency.

Persons on premises trained in first-aid:
Person to whom accidents must be reported:
Other information:

Logarithms

	0	1	2	3	4	5	6	7	8	9	Mean differences 1	2	3	4	5	6	7	8	9
1·0	0 0000	0043	0086	0128	0170	0212	0253	0294	0334	0374	4	8	12	17	21	25	29	33	37
1·1	0 0414	0453	0492	0531	0569	0607	0645	0682	0719	0755	4	8	11	15	19	23	27	30	34
1·2	0 0792	0828	0864	0899	0934	0969	1004	1038	1072	1106	3	7	10	14	17	21	24	28	31
1·3	0 1139	1173	1206	1239	1271	1303	1335	1367	1399	1430	3	6	10	13	16	19	23	26	29
1·4	0 1461	1492	1523	1553	1584	1614	1644	1673	1703	1732	3	6	9	12	15	18	21	24	27
1·5	0 1761	1790	1818	1847	1875	1903	1931	1959	1987	2014	3	6	8	11	14	17	20	22	25
1·6	0 2041	2068	2095	2122	2148	2175	2201	2227	2253	2279	3	5	8	11	13	16	19	21	24
1·7	0 2304	2330	2355	2380	2405	2430	2455	2480	2504	2529	3	5	8	10	13	15	18	20	23
1·8	0 2553	2577	2601	2625	2648	2672	2695	2718	2742	2765	2	5	7	9	12	14	16	19	21
1·9	0 2788	2810	2833	2856	2878	2900	2923	2945	2967	2989	2	4	7	9	11	13	16	18	20
2·0	0 3010	3032	3054	3075	3096	3118	3139	3160	3181	3201	2	4	6	8	11	13	15	17	19
2·1	0 3222	3243	3263	3284	3304	3324	3345	3365	3385	3404	2	4	6	8	10	12	14	16	18
2·2	0 3424	3444	3464	3483	3502	3522	3541	3560	3579	3598	2	4	6	8	10	12	14	15	17
2·3	0 3617	3636	3655	3674	3692	3711	3729	3747	3766	3784	2	4	6	7	9	11	13	15	17
2·4	0 3802	3820	3838	3856	3874	3892	3909	3927	3945	3962	2	4	5	7	9	11	12	14	16
2·5	0 3979	3997	4014	4031	4048	4065	4082	4099	4116	4133	2	3	5	7	9	10	12	14	15
2·6	0 4150	4166	4183	4200	4216	4232	4249	4265	4281	4298	2	3	5	7	8	10	12	13	15
2·7	0 4314	4330	4346	4362	4378	4393	4409	4425	4440	4456	2	3	5	6	8	9	11	13	14
2·8	0 4472	4487	4502	4518	4533	4548	4564	4579	4594	4609	2	3	5	6	8	9	11	12	14
2·9	0 4624	4639	4654	4669	4683	4698	4713	4728	4742	4757	1	3	4	6	7	9	10	12	13
3·0	0 4771	4786	4800	4814	4829	4843	4857	4871	4886	4900	1	3	4	6	7	9	10	11	13
3·1	0 4914	4928	4942	4955	4969	4983	4997	5011	5024	5038	1	3	4	6	7	8	10	11	12
3·2	0 5051	5065	5079	5092	5105	5119	5132	5145	5159	5172	1	3	4	5	7	8	9	11	12
3·3	0 5185	5198	5211	5224	5237	5250	5263	5276	5289	5302	1	3	4	5	7	8	9	10	12
3·4	0 5315	5328	5340	5353	5366	5378	5391	5403	5416	5428	1	3	4	5	6	8	9	10	11
3·5	0 5441	5453	5465	5478	5490	5502	5514	5527	5539	5551	1	2	4	5	6	7	9	10	11
3·6	0 5563	5575	5587	5599	5611	5623	5635	5647	5658	5670	1	2	4	5	6	7	8	10	11
3·7	0 5682	5694	5705	5717	5729	5740	5752	5763	5775	5786	1	2	3	5	6	7	8	9	10
3·8	0 5798	5809	5821	5832	5843	5855	5866	5877	5888	5899	1	2	3	4	6	7	8	9	10
3·9	0 5911	5922	5933	5944	5955	5966	5977	5988	5999	6010	1	2	3	4	6	7	8	9	10
4·0	0 6021	6031	6042	6053	6064	6075	6085	6096	6107	6117	1	2	3	4	5	6	7	9	10
4·1	0 6128	6138	6149	6160	6170	6180	6191	6201	6212	6222	1	2	3	4	5	6	7	8	9
4·2	0 6232	6243	6253	6263	6274	6284	6294	6304	6314	6325	1	2	3	4	5	6	7	8	9
4·3	0 6335	6345	6355	6365	6375	6385	6395	6405	6415	6425	1	2	3	4	5	6	7	8	9
4·4	0 6435	6444	6454	6464	6474	6484	6493	6503	6513	6522	1	2	3	4	5	6	7	8	9
4·5	0 6532	6542	6551	6561	6571	6580	6590	6599	6609	6618	1	2	3	4	5	6	7	8	9
4·6	0 6628	6637	6646	6656	6665	6675	6684	6693	6702	6712	1	2	3	4	5	6	7	7	8
4·7	0 6721	6730	6739	6749	6758	6767	6776	6785	6794	6803	1	2	3	4	5	5	6	7	8
4·8	0 6812	6821	6830	6839	6848	6857	6866	6875	6884	6893	1	2	3	4	5	5	6	7	8
4·9	0 6902	6911	6920	6928	6937	6946	6955	6964	6972	6981	1	2	3	4	4	5	6	7	8
5·0	0 6990	6998	7007	7016	7024	7033	7042	7050	7059	7067	1	2	3	3	4	5	6	7	8
5·1	0 7076	7084	7093	7101	7110	7118	7126	7135	7143	7152	1	2	3	3	4	5	6	7	8
5·2	0 7160	7168	7177	7185	7193	7202	7210	7218	7226	7235	1	2	3	3	4	5	6	7	8
5·3	0 7243	7251	7259	7267	7275	7284	7292	7300	7308	7316	1	2	2	3	4	5	6	6	7
5·4	0 7324	7332	7340	7348	7356	7364	7372	7380	7388	7396	1	2	2	3	4	5	6	6	7
	0	1	2	3	4	5	6	7	8	9	1	2	3	4	5	6	7	8	9

	0	1	2	3	4	5	6	7	8	9	Mean differences 1	2	3	4	5	6	7	8	9
5·5	0 7404	7412	7419	7427	7435	7443	7451	7459	7466	7474	1	2	2	3	4	5	5	6	7
5·6	0 7482	7490	7497	7505	7513	7520	7528	7536	7543	7551	1	2	2	3	4	5	5	6	7
5·7	0 7559	7566	7574	7582	7589	7597	7604	7612	7619	7627	1	2	2	3	4	5	5	6	7
5·8	0 7634	7642	7649	7657	7664	7672	7679	7686	7694	7701	1	1	2	3	4	4	5	6	7
5·9	0 7709	7716	7723	7731	7738	7745	7752	7760	7767	7774	1	1	2	3	4	4	5	6	7
6·0	0 7782	7789	7796	7803	7810	7818	7825	7832	7839	7846	1	1	2	3	4	4	5	6	6
6·1	0 7853	7860	7868	7875	7882	7889	7896	7903	7910	7917	1	1	2	3	4	4	5	6	6
6·2	0 7924	7931	7938	7945	7952	7959	7966	7973	7980	7987	1	1	2	3	4	4	5	6	6
6·3	0 7993	8000	8007	8014	8021	8028	8035	8041	8048	8055	1	1	2	3	3	4	5	6	6
6·4	0 8062	8069	8075	8082	8089	8096	8102	8109	8116	8122	1	1	2	3	3	4	5	5	6
6·5	0 8129	8136	8142	8149	8156	8162	8169	8176	8182	8189	1	1	2	3	3	4	5	5	6
6·6	0 8195	8202	8209	8215	8222	8228	8235	8241	8248	8254	1	1	2	3	3	4	5	5	6
6·7	0 8261	8267	8274	8280	8287	8293	8299	8306	8312	8319	1	1	2	3	3	4	5	5	6
6·8	0 8325	8331	8338	8344	8351	8357	8363	8370	8376	8382	1	1	2	3	3	4	4	5	6
6·9	0 8388	8395	8401	8407	8414	8420	8426	8432	8439	8445	1	1	2	3	3	4	4	5	6
7·0	0 8451	8457	8463	8470	8476	8482	8488	8494	8500	8506	1	1	2	2	3	4	4	5	6
7·1	0 8513	8519	8525	8531	8537	8543	8549	8555	8561	8567	1	1	2	2	3	4	4	5	5
7·2	0 8573	8579	8585	8591	8597	8603	8609	8615	8621	8627	1	1	2	2	3	4	4	5	5
7·3	0 8633	8639	8645	8651	8657	8663	8669	8675	8681	8686	1	1	2	2	3	4	4	5	5
7·4	0 8692	8698	8704	8710	8716	8722	8727	8733	8739	8745	1	1	2	2	3	4	4	5	5
7·5	0 8751	8756	8762	8768	8774	8779	8785	8791	8797	8802	1	1	2	2	3	3	4	5	5
7·6	0 8808	8814	8820	8825	8831	8837	8842	8848	8854	8859	1	1	2	2	3	3	4	5	6
7·7	0 8865	8871	8876	8882	8887	8893	8899	8904	8910	8915	1	1	2	2	3	3	4	4	5
7·8	0 8921	8927	8932	8938	8943	8949	8954	8960	8965	8971	1	1	2	2	3	3	4	4	5
7·9	0 8976	8982	8987	8993	8998	9004	9009	9015	9020	9025	1	1	2	2	3	3	4	4	5
8·0	0 9031	9036	9042	9047	9053	9058	9063	9069	9074	9079	1	1	2	2	3	3	4	4	5
8·1	0 9085	9090	9096	9101	9106	9112	9117	9122	9128	9133	1	1	2	2	3	3	4	4	5
8·2	0 9138	9143	9149	9154	9159	9165	9170	9175	9180	9186	1	1	2	2	3	3	4	4	5
8·3	0 9191	9196	9201	9206	9212	9217	9222	9227	9232	9238	1	1	2	2	3	3	4	4	5
8·4	0 9243	9248	9253	9258	9263	9269	9274	9279	9284	9289	1	1	2	2	3	3	4	4	5
8·5	0 9294	9299	9304	9309	9315	9320	9325	9330	9335	9340	1	1	2	2	3	3	4	4	5
8·6	0 9345	9350	9355	9360	9365	9370	9375	9380	9385	9390	1	1	2	2	3	3	4	4	5
8·7	0 9395	9400	9405	9410	9415	9420	9425	9430	9435	9440	1	1	2	2	3	3	4	4	5
8·8	0 9445	9450	9455	9460	9465	9469	9474	9479	9484	9489	0	1	1	2	2	3	3	4	4
8·9	0 9494	9499	9504	9509	9513	9518	9523	9528	9533	9538	0	1	1	2	2	3	3	4	4
9·0	0 9542	9547	9552	9557	9562	9566	9571	9576	9581	9586	0	1	1	2	2	3	3	4	4
9·1	0 9590	9595	9600	9605	9609	9614	9619	9624	9628	9633	0	1	1	2	2	3	3	4	4
9·2	0 9638	9643	9647	9652	9657	9661	9666	9671	9675	9680	0	1	1	2	2	3	3	4	4
9·3	0 9685	9689	9694	9699	9703	9708	9713	9717	9722	9727	0	1	1	2	2	3	3	4	4
9·4	0 9731	9736	9741	9745	9750	9754	9759	9763	9768	9773	0	1	1	2	2	3	3	4	4
9·5	0 9777	9782	9786	9791	9795	9800	9805	9809	9814	9818	0	1	1	2	2	3	3	4	4
9·6	0 9823	9827	9832	9836	9841	9845	9850	9854	9859	9863	0	1	1	2	2	3	3	4	4
9·7	0 9868	9872	9877	9881	9886	9890	9894	9899	9903	9908	0	1	1	2	2	3	3	4	4
9·8	0 9912	9917	9921	9926	9930	9934	9939	9943	9948	9952	0	1	1	2	2	3	3	4	4
9·9	0 9956	9961	9965	9969	9974	9978	9983	9987	9991	9996	0	1	1	2	2	3	3	4	4
	0	1	2	3	4	5	6	7	8	9	1	2	3	4	5	6	7	8	9

Reproduced from: Blewett and Hewstone, *Cassell's Four-Figure Tables*, Cassell, London, 3rd edn., 1979.

Answers to questions

Section 1

7(a) I = W/V = 5000/240 = 20.8 A; (b) 750/240 = 3.1 A.
26(a) Total power of appliances = 50 + 1000 + 500 + 250 + 100 = 1900W. Watts, W = Volts, V × amps, I; therefore current drawn through the cable at mains voltage (240V) = 1900/240 = 7.9A. It would be safe to operate these appliances as this current is within the 10A rating of the cable. The plug should be fitted with a 10A (or 13A) fuse.

(b) Total power of appliances = 1000 + 2000 + 500 + 100 + 750 = 4350W. Current required = W/V = 4350/240 = 18.1A. This current greatly exceeds the 10A cable rating and would produce serious overheating. It would be unsafe to connect the multiple socket outlet panel to the mains irrespective of the plug fuse rating.

Section 9

Fig. 9.6 vernier readings: (a) 19.7 mm, (b) 44.6 mm.
Fig. 9.10 micrometer readings:

(a)	Sleeve reading	= 0.20	cm
	Thimble reading	= 0.009	cm
	Total	= 0.209	cm
(b)	0.074 cm = 0.74 mm.		
(c)	Sleeve reading	= 0.25	cm
	Thimble reading	= 0.031	cm
	Vernier reading	= 0.0003	cm
	Total	= 0.2813	cm

Section 10

7.(a) 1.81×10^2. This has one more significant figure than 5.2 which is the limiting value, but expressing the answer as 1.8×10^2 would reduce the precision from 1 part in 52 (approx. 2%) to 1 part in 18 (approx. 5.6%).

(b) 1.00. All the figures do cancel out, but it would be very imprecise to give the answer as 1 as this would imply an uncertainty of ±1 or 100%. The limiting value is 1.01 in the calculation, so the result should be expressed to the same number of significant figures.

(c) 3.72×10^6.

8.(a) l = 1 m = 100 cm; t = 2.00 s. Uncertainty in l = 0.1 cm in 100 cm = 0.1%. Uncertainty in t = 0.1s in 2.00s = 5%. But time appears as a squared term in the expression for g. Therefore, uncertainty in $t^2 = 2 \times 5$ = 10% and maximum expected error in g = 0.1 + (2 × 5) = 10.1%.

(b) Uncertainty in l = 0.1 cm in 10 cm = 1%. Uncertainty in t = 0.1s in 0.63s = 15.9%; uncertainty in t^2 = 2 × 15.9 = 31.8%. Maximum expected error in g = 1 + 31.8 = 32.8%.

Answers to objective test

1. *E*.
2. *B* and *D*. These two methods would connect the live terminal to the earth lead and the operator could be electrocuted as soon as the equipment is plugged into a mains socket as the metal casing of the appliance would then be live.
3. *A* Current (I) = Watts (W)/Volts (V) $\therefore I =$ 5000/250 = 20 A.
4. *C* I = 2000/240 = 8.3 A and a 13 A fuse would be the most appropriate.
5. *A*.
6. *C*.
7. *A*.
8. *D*.
9. *F*.
10. *A*.
11. *B*.
12. *D*.
13. *C*.
14. *E*.
15. *D*.
16. *A*.
17. *D*.
18. *B* $\log y = \log k + n \log x$. The gradient, n, = 2 and the intercept on the $\log y$ axis when $\log x$ = O is also 2. If $\log k = 2$, $k = 10^2 = 100$.
19. *B*.
20. *C*.
21. *B*.
22. *D*.
23. *C*.
24. *A*.
25. *D*.
26. *D*.
27. *E*.
28. *A*.
29. *C*.
30. True, but he would be wise to have it checked before plugging it into a mains socket.
31. False.
32. True.
33. False.
34. True.
35. False.
36. True.
37. True.
38. True.
39. False.
40. True.
41. False.
42. True.
43. True.
44. False.
45. True.
46. False.
47. True.
48. True.
49. True.
50. True.
51. False.
52. True.
53. True.
54. True.
55. False.
56. False.
57. False.
58. True.
59. False. This statement applies to Class A burettes.
60. False.
61. False.
62. False. Mercury levels are read at the top of the meniscus; all other liquids

are read at the bottom of the meniscus (see fig. 9.18).

63. False.
64. True.
65. False. This statement applies to secondary cells.
66. False. The Act expressly forbids this.
67. False. He should be placed in the recovery position (see fig. 12.2)
68. False.
69. False.
70. False. A vein carries blood back to the heart. Blood loss is controlled by applying pressure at a suitable point on the artery which brings blood from the heart to the wound.
71. True. Rings should be removed before swelling occurs.
72. False and some highly toxic gases (e.g. carbon monoxide) do not have any smell.
73. True. He will almost certainly require an anaesthetic as soon as he reaches hospital and must not be given anything to eat or drink.
74. True.
75. False.

Index

Absorption spectrum, didymium, 136; polystyrene, 139
Accident reporting, 206
Accuracy and precision, 103
Acids, treatment of burns, 199
Alkali, treatment of burns, 199
Alkaline cells, 166, 169; charging, 170
Alpha rays, 48
Ampere, 2, 220
Ampere-hour, 167
Animals, 41
Antidote, acid burns, 199; alkali burns, 199; bromine, 199; cyanide poisoning, 195; phenol, 199; phosphorus, 199
Artery, 181
Artificial respiration, 180 *et seq*; purpose of, 182
Asbestos, 46, 66
Asphyxia, 174, 181
Asphyxiation, 72
Assembly points, 75
Automatic sprinkler system, 89

Back injuries, 90
Bacteriological cultures, disposal, 98; sterlisation of apparatus, 56; use of transfer chamber, 55, 85
Balances, 112 *et seq*; accuracy, 115; checking, 116; procedure for use, 113
Bandaging, 188 *et seq*
Bar chart, 156, 158, 159
Battery acid, preparation 167; specific density, 167, 169
BCF extinguisher, 66; operation, 73
Becquerel, 49
Benzene, 30 – 32
Beta rays, 49
Binocular microscope, 132
Biological laboratories, 55 *et seq*, 90; hazards of, 55, 56
Bleeding, control of, 190
Block diagram, 156
Blood, 180
Board of Trade unit, 8
Boring corks or rubber bungs, 45
Bottle carrier, 37
Breakages, reporting, 164
British Standards Institution (BSI), 4, 20, 25, 54, 78, 105, 123, 124, 135, 140 *et seq*; kitemark, 21, 140; purpose of, 140; safety mark, 140; titles of useful publications, 141, 142; Yearbook, 138, 141
Bromine, 199
Bruise, 187
Bung, boring, 45; insertion of glass tubing, 43 *et seq*
Burette, 117 *et seq*; accuracy, 123; cleaning, 120, 162; procedure for use, 118
Burning clothing, 74, 198
Burns, 175; chemical, 198 *et seq*; estimation of surface area, 197; treatment 197 *et seq*

Callipers, vernier, 105, 107; accuracy, 111
Candela, 221
Carbon dioxide extinguisher, 65; operation 73
Carbon tetrachloride, dangers of, 32; fire extinguisher, 66 *et seq*
Carbonyl chloride, 66, 88
Carboy 37; siphon, 37
Carcinogen, 23, 32, 45, 46
Cartridge fuse, colour code, 11; selection for stated application, 8, 11; testing for continuity, 9, 10
Caustic alkali, treatment of burns, 199
Cells, 164 *et seq*; capacity, 167; charging, 166, 170; discharge, 166; electrolyte, 167; reactions, 171
Centrifuges, hazards of, 27
Charging batteries, 170
Chemical burns, specific antidotes, 199; treatment, 198 *et seq*
Chemical laboratories, hazards of, 23
Chlorine, 25, 32
Circular diagram, 157, 159
Classification of fires, 60
Cleaning agents, 162 *et seq*
Cleaning, glassware, 162 *et seq*; microscopes, 133; volumetric glassware, 118, 120, 123, 162
Clothing, burning, 198
Code of behaviour, 87
Colour blindness, 3, 78
Colour code, cable and flex, 5; cartridge fuses, 11; fire extinguishers, 64 *et seq*; first aid boxes, 78, 176; gas cylinders, 25; resistors, 4
Coma; 203
Compressed gases, hazards of, 24, 25

Conductivity, 2
Connection of plug to flex end, 5
Containers, labelling, 38 *et seq*; storage, 31, 35, 36
Contusion, 187
Corks, boring, 45; insertion of glass tubing, 43 *et seq*; rolling, 43
Cork roller, 44
Crepitus, 205
Cumulative poisons, 31
Curie, 49
Current, calculation, 7; danger of drawing excessive, 16; definition, 1; effect on human body, 14; units, 2, 220
Cuts, 174, 187
Cutting tools, 41
Cyanosis, 181
Cylinder, clamp, 25, 26; colour code for, 25; hazards of, 24, 25, 27; trolley, 27

Dermatitis, 23, 31, 33
Diaphragm, 180
Dilated pupils, 185
Disposal, acids, 98; bacteriological cultures, 98; calcium dicarbide, 98; flammable liquids, 97; poisonous waste, 96; radioactive materials, 97; sodium or potassium, 98; regulations, 95 *et seq*
Distribution board, 17
Double insulation, 16
Dressings, 187

Earth, 2, 6
Earthing electrical equipment, 14, 15
Eating and drinking in the laboratory, 56, 88
Electrical hazards, 3, 13, 15, 16
Electrical safety, 3 *et seq*
Electricity, effect on human body, 14
Electric shock, 14, 176, 182; first aid treatment, 201; secondary injuries, 14
Electrocution, 14, 80, 182
Electromagnetic spectrum, 52
Errors, 147
Escape routes, 75
Ether (*see* ethoxyethane); peroxide formation in, 33
Ethoxyethane, dangers of, 33, 97
European standards, 141
Evacuated glassware, dangers of, 24
Experimental results, presentation, 149 *et seq*; reporting, 148
Explosive limits, 61, 62
Explosive substance, hazard warning symbol, 82
Eye flash, 53
Eye protection, 22, 53, 54
Eye splashes of chemical in, 175, 199
Eye wash bottle, siting, 176; use, 200

Fire alarm buttons, position of, 75
Fire blanket, 68; use, 74
Fires, classification of, 60
Fire damage, cost of, 59
Fire, definition, 59
Fire doors, purpose of, 63, 75
Foam extinguisher, 66; operation, 73
Fire extinguishers, colour coding, 64, 65, 66, 68; main types, 64 *et seq*; range of application, 69 *et seq*; use, 72 *et seq*
Fire fighting, 71 *et seq*
Fire hazards, 59
Fires, spread of, 63
Fire triangle, 58
First aid, 174 *et seq*; definition, 174; priorities in administering, 179; proficiency certificate, 174
First aid boxes, colour coding, 78; contents, 177, 178; siting, 176
Flammable and inflammable, meaning, 61
Flammable gas, 60, 226
Flammable liquids, 32; disposal, 97
Flammable solid, 61, 226
Flash point, 61, 62
Flex, colour code, 5; connection to three-pin plug top, 5; danger of damaged or carelessly routed, 16; rating, 12; selection for specified application, 12
Formalin, dangers of, 32;
Fracture, 176, 204; greenstick, 204; symptoms, 204; treatment, 205 *et sea*
Fume cupboard, 27 *et seq;* recommended entry velocity, 28, 29
Fuse, function of, 8; selection of appropriate, 8

Gamma rays, 49
G-clamp, 110
Glass tubing, cutting, 41; insertion into cork, 43, 44
Glassware, cleaning, 162 *et seq*; hazards of damaged, 41
Graphs, 150; straight-line, 150 *et seq*
Gray, 49

Hazards, biological and medical labora-

tories, 55, 56; broken glassware, 41; centrifuges, 27; chemical laboratories, 23 *et seq*; compressed air, 24; eating and drinking in laboratories, 56, 88; electrical equipment, 3 *et seq*; evacuated glassware, 24; fire, 58 *et seq*; gas cylinders, 25, 26, 27; lasers, 54, 79; long hair, 21; loose clothing, 21; mixtures of chemicals, 34, 35; oxygen cylinders, 27; physical laboratories, 48 *et seq*; power tools, 20 *et seq*; radioactive sources, 48, 79; smoking in a laboratory, 88
Hazardous mixtures and reactions, 34, 35
Hazard warning symbols, 39, 40, 78 *et seq*, 226; biological hazard, 81; chemical hazards, 82, 83, 226; high voltage, 80; ionising radiation, 79; lasers, 79; radiation, 79; radioactive substances, 83
Hazardous reactions, 34 *et seq*
Hazchem code, 226
Health and Safety at Work, etc., Act, 95, 99, 100
Heart massage, 185
Hernia, 90, 91
High voltage warning symbol, 80
Histogram, 157
Horsepower, 8, 223
Hydrometer, 167
Hygiene, 41

Ignition temperature, 61, 62
Implosion, 24, 25
Inflammable and flammable, meaning, 61
Injuries in laboratory accidents, 174 *et seq*
Intervertebral discs, 91, 92
Isolation box, 51

Kilogram, 134
Kilowatt-hour, 8
Kiss of life, 183
Kitemark, 21, 140

Lacerations, 174, 187
Laser light, 54; hazards of, 54; hazard warning sign, 79; safety procedure, 54;
Law, 95 *et seq*
Lead-acid battery, 166, 168; charging, 170; electrolyte, 167; reactions, 171
Lifting, correct procedure for, 92, 93
Live terminal, 2, 6
Long hair, hazards of, 21
Loose clothing, hazards of, 21

Magnification of microscope, 130, 133
Mains voltage, 2; dangers of, 13
Mallory-Ruben mercury cell, 166
Manipulator glove box, 51
Massage, heart, 185
Mercury, 30, 31; cell, 166
Methanal (formaldehyde); 32, 46
Methanol, dangers of, 33
Metallurgical microscope, 128 *et seq*
Metre, 134
Micrometer, 107 *et seq*; accuracy, 111
Microbiological equipment, sterilisation, 56
Microbiological experiments, 56, 85
Micro-organisms, 55, 56, 85
Microscope, 126 *et seq*; cleaning, 133; magnification, 130, 133; parts of, 126; procedure for use, 131; ray diagram, 127; reflecting, 128, 130
Multimeter, 124 *et seq*; for checking continuity of cartridge fuse, 9
Multiple adaptors, 17

Neutral, 2, 6
Nickel-cadmium cell, 166, 169; charging, 170
Nickel-iron (NiFe) cell, 166, 169; charging, 170; reactions, 171
NiFe cell, 166, 169

Objective, 126, 128 *et seq*
Ohm, 2
Ohm's Law, 3, 7, 12, 13, 80
Oil immersion objective, 128, 130
Overloaded socket, danger of, 18

Parallax, 9, 106, 118, 199, 125
Pathogenic organisms, 56
Penetration of radiation, 48, 49
Peroxide formation in ethers, 33
pH, 135, 136
Phenol, 199
Phosgene, 66, 88
Phosphorus, 199
Physiological effect of electric current, 14
Pictogram, 158, 160
Pictorial bar chart, 158, 160
Pie chart, 157, 159
Pipette, 121 *et seq*; accuracy, 123; clean-

ing, 123, 162; procedure for use, 121
Plug, connection to a flex end, 5
Poisons, antidotes, 195, 199; general treatment for, 194; ingestion, 195, inhalation, 195; injection, 196; labelling, 40; procedure when working with, 31; skin absorption, 196; storage, 29, 31
Poisoning, 175
Poisonous waste, disposal, 96
Potential (or voltage), units, 2
Potassium, disposal, 98; fires 69, 71
Powders, as a fire hazard, 59
Powder extinguisher, 67; operation, 73
Power, 3, 7
Power tools, 20 *et seq*
P.p.m. (or parts per million), 30
Practical notebook, 145
Precision and accuracy, 103
Pressure points, definition, 192; location, 193; use, 192 *et seq*
Primary standards, 134 *et seq*
Protective clothing, 21, 51, 53, 54, 85, 86
Pupils, dilated, 185

Rad, 49
Radiation, detection, 48, 50; doses, 49, 50; effect on human body, 48 *et seq*; hazard warning symbol, 79, 83; types of, 48; units, 49
Radioactive materials, disposal, 97
Radioactive sources, legal restrictions, 51; recommendations when using, 51; storage, 52
Radioactivity, 48 *et seq*
Recovery position, 179
Rem, 50
Respiration, artificial, 183; mechanism of, 182
Respiratory system, 180
Reporting, accident, 206; scientific, 145 *et seq*
Resistance, units, 2
Rolling corks, 43
Rubber gloves, 86

Safety bulb, 122
Safety, electrical, 3 *et seq*
Safety glass in fume-cupboards, 29
Safety goggles, 85
Safety mark, 140
Safety officer, functions of, 100
Safety spectacles, 85
Sand, 67
Scalds, 175
Scientific reporting, 145 *et seq*
Second, 134
Secondary cells, 164 *et seq*
Secondary standard, 134 *et seq*
Shock, electric, 13 *et seq*, 20, 201; wound, 202
Sievert, 50
Significant figures, 146
Siphon, 37, 38
SI units, 220 *et seq*
Skin, electrical resistance of, 13, 80; splashes of chemical on, 176
Smoke doors, 63, 75
Smoking in a laboratory, 88
Socket, danger of damaged, 16; danger of overloaded, 18; position of terminals in, 2
Soda-acid extinguisher, 64
Sodium, disposal, 98; fires, 69, 71
Solvents, disposal, 97; storage, 35
Solvent store, 35
Specific gravity, 167
Spine, 91, 92
Standard cell, 125
Standard gauge block, 111, 135
Standard gauze dressing, 187
Standards, 134 *et seq;* concentration, 137, 227; conductivity, 137, 226; disintegration rate; 135; length, 134; mass, 134; optical density (absorbance), 137, 227; pH, 135; primary, 134 *et seq*; secondary, 134 *et seq*; time, 134; volume, 138; voltage, 135, 136; wave length, 137; wave number, 137;
Standard weights, 116
Sterilisation of microbiological equipment, 56
Storage, of acids, 36; flammable liquids, 35; poisons, 31
Straight line graphs, 150 *et seq*
Sulphuric acid, dilution, 35, 167; hazards, 35

Tables, 149 *et seq*
Teratogens, 46
Terminals, position in three-pin plug top, 6; position in three-pin socket outlet, 2
Testing, of instruments, 116, 134 *et seq*, 138
Tetrachloromethane (carbon tetrachloride), dangers of, 32, 66; fire extinguisher, 66 *et seq*

Thermal burns, estimation of surface area, 197; treatment 197 *et seq*
Three-pin socket outlet and plugs, 2, 6
Threshold limit value (TLV), 29
Trachea, 180
Transfer chamber, 56, 85

Ultra-violet radiation, 52, 53; protection against hazards of, 53
Unconsciousness, 175, 181; treatment, 182, 183, 203
Unit, Board of Trade, 8

Valve stem, 26
Vaporising liquid extinguishers, 66 *et seq*; operation, 73
Vein, 181
Vernier, 103 *et seq*
Vernier callipers, 105, 107; accuracy, 111
Vernier microscope, 106
Vertebra, 91, 92
Volt, 2
Voltage drop, 12
Volumetric glassware, 118 *et seq*; accuracy, 123, 124; cleaning, 118, 162; markings, 118

Warning symbols, 39, 40, 79 *et seq*, 226; biohazard, 81; chemical hazards, 82, 226; high voltage, 80; ionising radiation, 79; lasers, 79; radioactive substances, 83; radioactivity, 79;
Water extinguisher, 64; operation, 73
Watt, 2, 7
Weighing, procedure for, 113
Weights, standard, 116
Weston standard cell, 125, 135, 136
Winchester carrier, 37
Windpipe, 180
Wire strippers, 5
Wiring a 13-amp plug, 5
Wounds, classification, 186 *et seq*
Wound shock, symptoms, 202; treatment, 202

X-rays, 49 *et seq*